TUMORS
and
CANCERS
CENTRAL AND PERIPHERAL
NERVOUS SYSTEMS

T0100354

POCKET GUIDES TO
BIOMEDICAL SCIENCES

https://www.crcpress.com/Pocket-Guides-to-Biomedical-Sciences/book-series/
CRCPOCGUITOB

The *Pocket Guides to Biomedical Sciences* series is designed to provide a concise, state-of-the-art, and authoritative coverage on topics that are of interest to undergraduate and graduate students of biomedical majors, health professionals with limited time to conduct their own searches, and the general public who are seeking for reliable, trustworthy information in biomedical fields.

TUMORS
and
CANCERS
CENTRAL AND PERIPHERAL
NERVOUS SYSTEMS

Dongyou Liu

CRC Press
Taylor & Francis Group
Boca Raton London New York

CRC Press is an imprint of the
Taylor & Francis Group, an **informa** business

CRC Press
Taylor & Francis Group
6000 Broken Sound Parkway NW, Suite 300
Boca Raton, FL 33487-2742

© 2018 by Taylor & Francis Group, LLC
CRC Press is an imprint of Taylor & Francis Group, an Informa business

No claim to original U.S. Government works

Printed on acid-free paper

International Standard Book Number-13: 978-1-4987-2969-7 (Paperback)
978-1-1383-0020-0 (Hardback)

Library of Congress Cataloging-in-Publication Data

Names: Liu, Dongyou, author.
Title: Tumors and cancers : central and peripheral nervous systems / Dongyou Liu.
Description: Boca Raton : Taylor & Francis, 2018. | Series: Pocket guides to biomedical sciences | "A CRC title, part of the Taylor & Francis imprint, a member of the Taylor & Francis Group, the academic division of T&F Informa plc." | Includes bibliographical references and index.
Identifiers: LCCN 2017007750 | ISBN 9781498729697 (paperback : alk. paper) : 9781138300200 (hardback)
Subjects: LCSH: Brain--Cancer. | Central nervous system--Cancer.
Classification: LCC RC280.B7 L58 2018 | DDC 616.99/481--dc23
LC record available at https://lccn.loc.gov/2017007750

Visit the Taylor & Francis Web site at
http://www.taylorandfrancis.com

and the CRC Press Web site at
http://www.crcpress.com

Contents

Series Preface

Dramatic breakthroughs and nonstop discoveries have rendered biomedicine increasingly relevant to everyday life. Keeping pace with all these advances is a daunting task, even for active researchers. There is an obvious demand for succinct reviews and synthetic summaries of biomedical topics for graduate students, undergraduates, faculty, biomedical researchers, medical professionals, science policy makers, and the general public.

Recognizing this pressing need, CRC Press has established the *Pocket Guides to Biomedical Sciences* series, with the main goal to provide state-of-the-art, authoritative reviews of far-ranging subjects in short, readable formats intended for a broad audience. Volumes in the series will address and integrate the principles and concepts of the natural sciences and liberal arts, especially those relating to biomedicine and human well-being. Future volumes will come from biochemistry, bioethics, cell biology, genetics, immunology, microbiology, molecular biology, neuroscience, oncology, parasitology, pathology, and virology, as well as other related disciplines.

Representing one of the four volumes devoted to human tumors and cancers in the series, the current volume focuses on the central and peripheral nervous systems. Characterized by the uncontrolled growth of abnormal cells that often extend beyond their usual boundaries and disrupt the normal functions of affected organs, tumors and cancers are insidious diseases with serious consequences. Relative to our ongoing research and development efforts, our understanding of tumors and cancers remains rudimentary, and the arsenal at our disposal against these increasingly prevalent diseases are severely limited. The goal of this volume is the same as the goal for the series—to simplify, summarize, and synthesize a complex topic so that readers can reach to the core of the matter without the necessity of carrying out their own time-consuming literature searches.

We welcome suggestions and recommendations from readers and members of the biomedical community for future topics in the series and experts as potential volume authors/editors.

Dongyou Liu
Sydney, Australia

Contributors

Samer Hoz Saad Alameri
Arabic Board of Health
    Specialization Program
Department of Neurosurgery
Neurosurgery Teaching Hospital
Baghdad, Iraq

Mandana Behbahani
Division of Pediatric Neurosurgery
Ann and Robert H. Lurie Children's
    Hospital of Chicago
Chicago, Illinois

Stephen W. Coons
Department of Neuropathology
Barrow Neurological Institute
Phoenix, Arizona

Prantik Das
Northern Ireland Cancer Centre
Belfast, United Kingdom

Andreas K. Demetriades
Department of Neurosurgery
Western General Hospital
Edinburgh, UK

Mohamed Said Elsanafiry
Neurosurgery Department
Faculty of Medicine
Menoufia University
Menoufia, Egypt

Tayfun Hakan
Vocational School of Health Sciences
Okan University
İstanbul, Turkey

Mohammed Hmoud
Department of Clinical Affairs
College of Medicine
King Saud bin Abdulaziz University
    for Health Sciences
Jeddah, Saudi Arabia

Guoquan Jiang
Department of Neurosurgery
The Second Affiliated Hospital
Wannan Medical College
Wuhu, People's Republic of China

Christopher Kalhorn
Department of Neurosurgery
MedStar Georgetown University
    Hospital
Washington, D.C.

Divya Khosla
Department of Radiotherapy and
    Oncology
Government Medical College and
    Hospital
Chandigarh, India

Ritesh Kumar
Department of Radiotherapy
All India Institute of Medical
    Sciences
New Delhi, India

Victoria M. Lim
School of Medicine
Creighton University
Omaha, Nebraska

Dongyou Liu
Royal College of Pathologists of
   Australasia Quality Assurance
   Programs
Sydney, New South Wales
   Australia

Austin Huy Nguyen
School of Medicine
Creighton University
Omaha, Nebraska

Omekareswar Rambarki
Department of Neurosurgery
Nizam's Institute of Medical Sciences
Hyderabad, India

Alaa Samkari
Department of Pathology and
   Laboratory Medicine
King Abdullah International
   Medical Research Center
Jeddah, Saudi Arabia

Hasan R. Syed
Division of Pediatric Neurosurgery
Ann and Robert H. Lurie Children's
   Hospital of Chicago
Chicago, Illinois

Tadanori Tomita
Division of Pediatric Neurosurgery
Ann and Robert H. Lurie Children's
   Hospital of Chicago
Chicago, Illinois

Soubhagya Ranjan Tripathy
P.D. Hinduja Hospital and Medical
   Research Center
Mumbai, India

Adam M. Vaudreuil
School of Medicine
Creighton University
Omaha, Nebraska

Vamsi Krishna Yerramneni
Department of Neurosurgery
Nizam's Institute of Medical
   Sciences
Hyderabad, India

Xintong Zhao
Department of Neurosurgery
Yijishan Hospital
Wannan Medical College
Wuhu, People's Republic of China

1
Introductory Remarks

1.1 Preamble

Tumors and cancers are insidious diseases that result from uncontrolled growth of abnormal cells in one or more parts of the body.* Tumors and cancers have acquired a notorious reputation due not only to their ability to exploit host cellular machineries for their own advantage but also to their potential to cause human misery.

With a rapidly aging world population, widespread oncogenic viruses, and constant environmental pollution and destruction, tumors and cancers are poised to exert an increasingly severe toll on human health and well-being. There is a burgeoning interest from health professionals and the general public in learning about tumor and cancer mechanisms, clinical features, diagnosis, treatment, and prognosis. The following pages in the current volume as well as those in the sister volumes represent a concerted effort to satisfy this critical need.

1.2 Tumor mechanisms

The human body is composed of various types of cells that grow, divide, and die in an orderly fashion (so-called apoptosis). However, when some cells in the body change their growth patterns and fail to undergo apoptosis, they often produce a solid and sometimes nonsolid tumor (as in the blood). A tumor is considered benign if it grows but does not spread beyond the immediate area in which it arises. While most benign tumors are not life-threatening, those found in the brain can be deadly. In addition, some benign tumors are precancerous, with the propensity to become cancer if left untreated. By contrast, a tumor is considered malignant and cancerous if it grows continuously and spreads to surrounding areas and other parts of the body through the blood or lymph system.

A tumor located in its original (primary) site is known as a *primary tumor.* Cancer that spreads from its original (primary) site via the neighboring

---

* The terms *tumor* and *cancer*, along with *neoplasm* and *lesion*, are used interchangeably in colloquial language and publication. See Glossary.

tissue, the bloodstream, or the lymphatic system to another site of the body is called *metastatic cancer* (or *secondary cancer*). A metastatic cancer has the same name and the same type of cancer cells as the primary cancer. For instance, metastatic cancer in the brain that originates from breast cancer is known as *metastatic breast cancer*, not brain cancer.

Typically, tumors and cancers form in tissues after the cells undergo genetic mutations that lead to sequential changes known as *hyperplasia*, *metaplasia*, *dysplasia*, *neoplasia*, and *anaplasia* (see Glossary). Factors contributing to genetic mutations in the cells may be chemical (e.g., cigarette smoking, asbestos, paint, dye, bitumen, mineral oil, nickel, arsenic, aflatoxin, and wood dust), physical (e.g., sun, heat, radiation, and chronic trauma), viral (e.g., EBV, HBV, HPV, HTLV-1), immunological (e.g., AIDS and transplantation), endocrine (e.g., excessive endogenous or exogenous hormones), or hereditary (e.g., familial inherited disorders).

In essence, tumorigenesis is a cumulative process that demonstrates several notable hallmarks, including the following: (i) *sustaining proliferative signaling;* (ii) *activating local invasion and metastasis*; (iii) *resisting apoptosis and enabling replicative immortality*; (iv) *inducing angiogenesis* and *inflammation*; (v) *evading immune destruction*; (vi) *deregulating cellular energetics;* and (viii) *genome instability and mutation*.

1.3 Tumor classification, grading, and staging

Tumors and cancers are usually named for the organs or tissues in which they start (e.g., brain cancer, breast cancer, lung cancer, lymphoma, skin cancer, etc.). Depending on the type of tissue involved, tumors and cancers are grouped into a number of broad categories: (i) carcinoma (involving the epithelium); (ii) sarcoma (involving soft tissue); (iii) leukemia (involving blood-forming tissue); (iv) lymphoma (involving lymphocytes); (v) myeloma (involving plasma cells); (vi) melanoma (involving melanocytes); (vii) central nervous system (CNS) cancer (involving the brain or spinal cord); (viii) germ cell tumor (involving cells that give rise to sperm or eggs); (ix) neuroendocrine tumor (involving hormone-releasing cells); and (x) carcinoid tumor (a variant of neuroendocrine tumor found in the intestine).

The human nervous system consists of two main parts: the central nervous system (CNS) and the peripheral nervous system (PNS). The CNS includes the brain and spinal cord, while the PNS comprises the cranial nerves, spinal nerves (their roots and branches), peripheral nerves, and neuromuscular junctions. The nerves in the PNS link the CNS to sensory organs (e.g., the eye and ear), muscles, blood vessels, and glands, as well as other organs of the body.

Based on histopathological and clinical characteristics, major primary tumors affecting the CNS and PNS are divided into the following categories: (i) tumors of the neuroepithelial tissue (e.g., astrocytic tumors, other gliomas, oligodendroglial tumors, oligoastrocytic tumors, ependymal tumors, choroid plexus tumors, neuronal and mixed neuronal–glial tumors, pineal tumors, and embryonal tumors), (ii) tumors of the cranial and paraspinal nerves (e.g., neurofibroma, perineurioma, malignant peripheral nerve sheath tumor, and schwannoma), (iii) tumors of the meninges (e.g., meningiomas, melanocytic tumors, hemangiopericytoma, and hemangioblastoma), and (iv) tumors of the sellar region (e.g., craniopharyngioma) [1,2].

Under the auspice of the World Health Organization (WHO), the *International Classification of Diseases for Oncology, 3rd edition* (ICD-O-3) [3] has designed a five-digit system to classify tumors, with the first four digits being morphology code and the fifth digit being behavior code [1]. The fifth-digit behavior codes for neoplasms include the following: /0 benign; /1 benign or malignant; /2 carcinoma *in situ*; /3 malignant, primary site; /6 malignant, metastatic site; and /9 malignant, primary or metastatic site. For example, meningioma has an IDC-O code of 9530/0 and is a WHO Grade I tumor; atypical meningioma has an IDC-O code of 9539/1 and is a WHO Grade II tumor; and anaplastic (malignant) meningioma has an IDC-O code of 9530/3 and is a WHO Grade III tumor [1,3].

According to the amount of abnormality (including the level of differentiation from normal cells, the speed of growth, and the likelihood of spread), the 2016 *WHO Classification of Tumours of the Central Nervous System* separates CNS tumors into four grades: I (low grade, well differentiated, with the tumor appearing close to normal cells and tissue and growing slowly), II (intermediate grade, moderately differentiated), III (high grade, poorly differentiated), and IV (high grade, undifferentiated, with the tumor being unlike normal cells and tissue, growing rapidly, and spreading faster than lower grades) [1].

To further delineate tumors and cancers and assist in their treatment and prognosis, the stages of solid tumor are often determined by using the TNM system (see Glossary) of the American Joint Commission on Cancer (AJCC) according to the size and extent of the primary tumor (designated *T*, ranging from TX, T0, T1, T2, T3, to T4), the number of nearby lymph nodes involved (*N*, ranging from NX, N0, N1, N2, to N3), and the presence of distant metastasis (*M*, ranging from MX, M0, to M1) [4]. Therefore, under the TNM system, the pathological stage of a given tumor or cancer will be referred to as T1N0MX or T3N1M0 (with numbers after each letter giving more details about the tumor or cancer). However, for simplicity, five less-detailed TNM stages (0, I, II, III, and IV) are adopted clinically [4].

Another staging system that is more often used by cancer registries than by doctors divides tumors and cancers into five categories: (i) *in situ* (abnormal cells are present but have not spread to nearby tissue); (ii) localized (cancer is limited to the place where it started, with no sign that it has spread); (iii) regional (cancer has spread to nearby lymph nodes, tissues, or organs); (iv) distant (cancer has spread to distant parts of the body); and (v) unknown (there is not enough information to figure out the stage).

1.4 Tumor diagnosis

Because most tumors and cancers cause nonspecific, noncharacteristic clinical signs, a variety of procedures and tests are employed for their diagnosis.

Physical examination uncovers clinical signs of disease (e.g., lumps and other abnormalities).

Medical history review yields further clues to potential risk factors that enhance cancer development.

MRI (*magnetic resonance imaging*, also called *nuclear magnetic resonance imaging* or NMRI) utilizes a magnet, radio waves, and a computer to take detailed pictures of affected areas inside the body that help pinpoint the location and dimension of the tumor mass.

CT (*computed tomography*, also called *computerized tomography* or *computerized axial tomography*) scan links an X-ray machine to a computer together with a dye to take detailed pictures of affected organs or tissues in the body from different angles, revealing the precise location and dimension of the tumor.

PET (*positron emission tomography*) scan combines a computer-based scan with a radioactive glucose (sugar) to generate a rotating picture of the affected area, with malignant tumor cells showing up brighter due to their more active uptake of glucose than normal cells.

Ultrasound employs a device to deliver sound waves that bounce off tissues inside the body like an echo and a computer to record the echoes to create a picture (*sonogram*) of areas inside the body.

Endoscopy utilizes an endoscope (a thin, tube-like instrument with a light and a lens) to check for abnormal areas inside the body.

Biopsy procedures remove tumor and cancer cells or tissues for microscopic examination using hematoxylin and eosin or immunohistochemical stains.

This helps verify whether tumor and cancer cells are present at the edge of the material removed (positive margins) or not (negative, not involved, clear, or free margins), or whether they are neither negative nor positive (close margins).

Laboratory tests assess cerebrospinal fluid, blood, urine, or other body fluids for altered levels of substances in the body. These range from biochemical tests, fluorescence *in situ* hybridization, polymerase chain reaction, Southern and Western blot hybridizations, flow cytometry, and so on [2].

1.5 Tumor treatment and prognosis

Standard cancer treatments consist of surgery (for removal of tumor and relieving symptoms associated with tumor), radiotherapy (also called *radiation therapy* or *X-ray therapy*; delivered externally through the skin or internally [brachytherapy] for destruction of cancer cells or impeding their growth), chemotherapy (for inhibiting the growth of cancer cells, suppressing the body's hormone production or blocking the effect of the hormone on cancer cells, etc.; usually delivered via the bloodstream or by oral ingestion), and complementary therapies (for enhancing patients' quality of life and improving well-being). Depending on the circumstances, surgery may be used in combination with radiotherapy and/or chemotherapy to ensure that any cancer cells remaining in the body are eliminated.

The outcomes of tumor and cancer treatment include (i) cure (no traces of cancer remain after treatment and the cancer will never come back); (ii) remission (the signs and symptoms of cancer are reduced; in a complete remission, all signs and symptoms of cancer disappearing for 5 years or more suggests a cure); and (iii) recurrence (a benign or cancerous tumor comes back after surgical removal and adjunctive therapy).

Prognosis (or chance of recovery) is usually dependent on the location, type, and grade/stage of tumor, patient's age and health status, etc. Regardless of tumor/cancer types, patients with lower grade neoplasms generally have a better prognosis than those with higher grade neoplasms.

1.6 Future perspective

Tumors and cancers are a biologically complex disease that is expected to surpass heart disease in the coming decades to become the leading cause of death throughout the world. Despite extensive past research and development efforts, tumors and cancers remain poorly understood and effective cures remain largely elusive.

The completion of the Human Genome Project in 2003 and the establishment of the Cancer Genome Atlas in 2005 offered the promise of a better understanding of the genetic basis of human tumors and cancers and opened new avenues for developing novel diagnostic techniques and effective therapeutic measures.

Nonetheless, a multitude of factors pose continuing challenges for the ultimate conquest of tumors and cancers. These include the inherent biological complexity and heterogeneity of tumors and cancers, the contribution of various genetic and environmental risk factors, the absence of suitable models for human tumors and cancers, and the difficulty in identifying therapeutic compounds that kill or inhibit cancer cells only and not normal cells. Further effort is necessary to help overcome these obstacles and enhance the well-being of cancer sufferers.

Acknowledgments

Credits are due to a group of leading neuro-oncologists, whose expert contributions have greatly enriched this volume.

References

1. Louis DN, Perry A, Reifenberger G, et al. The 2016 World Health Organization classification of tumors of the central nervous system: A summary. *Acta Neuropathol* 2016; 131: 803–20.
2. Hoshide R, Jandial R. 2016 WHO classification of central nervous system tumors: An era of molecular biology. *World Neurosurg* 2016 Jul 28; pii: S1878–8750(16)30626-X.
3. Fritz A, et al. *International Classification of Diseases for Oncology.* Third edition. Geneva: World Health Organization, 2000.
4. Edge SB, Byrd DR, Compton CC, Fritz AG, Greene FL, Trotti A (editors). *AJCC Cancer Staging Manual.* Seventh edition. New York, NY: Springer, 2010.

SECTION I
Tumors of Neuroepithelial Tissue

2
Astrocytoma

2.1 Definition

Astrocytoma (or *astrocytic tumor*) represents a diverse group of brain tumors that arises from astrocytes (or astrocyte-like glial cells) in the central nervous system (CNS). Based on histopathological and molecular criteria, the 2016 WHO classification of CNS tumors separates astrocytoma into three categories: diffuse astrocytic tumors, other astrocytic tumors, and other gliomas [1].

Within the diffuse astrocytic tumor category, five types are recognized: (i) diffuse astrocytoma, isocitrate dehydrogenase (IDH) mutant (gemistocytic astrocytoma, IDH mutant)/diffuse astrocytoma, IDH wild-type/diffuse astrocytoma, not otherwise specified (NOS; WHO Grade II); (ii) anaplastic astrocytoma, IDH mutant/anaplastic astrocytoma, IDH wild-type/anaplastic astrocytoma, NOS (WHO Grade III); (iii) glioblastoma, IDH wild-type (giant cell glioblastoma/gliosarcoma/epithelioid glioblastoma)/glioblastoma, IDH mutant/glioblastoma, NOS (WHO Grade IV); and (iv) diffuse midline glioma, H3K27M mutant (WHO Grade IV).

Within the other astrocytic tumor category, four types are recognized: (i) pilocytic astrocytoma (PA; pilomyxoid astrocytoma, PMA) (WHO Grade I); (ii) subependymal giant cell astrocytoma (WHO Grade I); (iii) pleomorphic xanthoastrocytoma (WHO Grade II); and (iv) anaplastic pleomorphic xanthoastrocytoma (WHO Grade III).

Within the other glioma category, three types are recognized: (i) chordoid glioma of the third ventricle (WHO Grade I), (ii) angiocentric glioma (WHO Grade I), and (iii) astroblastoma (WHO Grade I) [1].

It should be noted that assignment of gliomas to WHO Grades (I, II, III, and IV) is largely based on the observation of certain histopathological and biological characteristics such as atypia, mitosis, endothelial proliferation, and necrosis (see the glossary). Gliomas displaying none of these features are regarded as Grade I, gliomas displaying one of these features (usually atypia) are regarded as Grade II, gliomas displaying two of these features are regarded as Grade III, and gliomas displaying three or four of these features are regarded as Grade IV [1].

2.2 Biology

Glial cells (also called *neuroglia* or simply *glia*) are non-neuronal cells in the brain and spinal cord of humans that support and protect neurons by forming myelin, removing dead neurons, and maintaining homeostasis in the central and peripheral nervous systems. Within the CNS, four types of glial cells can be identified: astrocytes, ependymal cells, microglia, and oligodendrocytes. Within the peripheral nervous system (PNS), two types of glial cells have been observed: Schwann cells and satellite cells.

Astrocytes (also called *astrocytic glial cells*, *macroglial cells*, or *astroglia*) are characteristic star-shaped glial cells evolving from heterogeneous populations of progenitor cells in the neuroepithelium of the developing CNS. The astrocyte precursors migrate to their final positions within the nervous system before the process of terminal differentiation occurs. Astrocytes may appear as either a fibrous (in white matter), protoplasmic (in gray matter), or radial (in planes perpendicular to the axes of ventricles) form in the brain and spinal cord.

Tumors arising from glial cells (including undifferentiated or partially differentiated astrocytes, oligodendrocytes, or ependymal cells) in the CNS are commonly referred to as *gliomas*. Astrocytomas, the most commonly reported gliomas, develop from astrocytes (or astrocyte-like glial cells). Occasionally, glial progenitors or neural stem cells may also give rise to astrocytomas. Astrocytomas occur in various parts of the brain and sometimes in the spinal cord but not usually beyond the brain and spinal cord.

According to the level of infiltration, astrocytoma may be divided into two broad classes: (i) that showing narrow zones of infiltration (e.g., PA, subependymal giant cell astrocytoma, pleomorphic xanthoastrocytoma), which is mostly noninvasive and clearly outlined on diagnostic images and (ii) that showing diffuse zones of infiltration (e.g., diffuse astrocytoma, anaplastic astrocytoma, glioblastoma), which demonstrates a preference for the cerebral hemispheres and an intrinsic tendency to progress to more advanced grades.

2.3 Epidemiology

Grade I astrocytomas constitute 2% of all CNS tumors, Grade II astrocytomas 8%, Grade III anaplastic astrocytomas 20%, and Grade IV glioblastomas 70%.

Astrocytomas can occur at any age, with Grades I and II (so-called low grade) tumors commonly found in children and Grades III and IV (so-called high grade) tumors prevalent in adults. The peak incidences of diffuse astrocytoma (Grade II), anaplastic astrocytoma (Grade III), and glioblastoma (Grade IV) are in the fourth, fifth, and seventh decades, respectively. There is a male predominance among astrocytoma-affected patients.

2.4 Pathogenesis

Astrocytomas induce regional effects through compression, invasion, and destruction of the brain parenchyma, arterial and venous hypoxia, competition for nutrients, release of metabolic end products (e.g., free radicals, altered electrolytes, neurotransmitters), and release and recruitment of cellular mediators (e.g., cytokines) that disrupt normal parenchymal function. Secondary clinical sequelae may result from elevated intracranial pressure caused by direct mass effect, increased blood volume, or increased cerebrospinal fluid volume.

Possible risk factors for astrocytoma range from past brain radiation therapy to genetic disorders (e.g., Li–Fraumeni syndrome, Turcot–Lynch syndrome, neurofibromatosis Type 1 and tuberous sclerosis). Molecularly, mutations in the *BRAF* gene are observed in PA and pleomorphic xanthoastrocytoma; alterations in IDH1, TP53, and ATRX are noted in Grades II and III astrocytoma, oligodendroglioma, and oligoastrocytoma, as well as secondary glioblastoma. There is evidence that astrocytomas may develop through IDH mutation followed by p53 mutation. These alterations influence cell metabolism, signal transduction pathways, global DNA methylation patterns, chromatin remodeling, and telomere length, leading to uncontrolled cell proliferation [2,3].

2.5 Clinical features

Clinical symptoms of astrocytoma include headache (which may disappear after vomiting); nausea; diplopia (double vision); dysphasia (speech problem); ataxia (loss of balance and trouble walking); hemiparesis (weakness or change in feeling on one side of the body); lethargy (unusual sleepiness or weakness); change in personality or behavior; unusual weight loss or gain; increase in the size of the head (in infants); and seizures.

In infants and young children, low-grade astrocytoma in the hypothalamus may lead to diencephalic syndrome (failure to thrive in emaciated, seemingly euphoric children; macrocephaly; intermittent lethargy; and visual impairment).

2.6 Diagnosis

Diagnosis of astrocytomas involves medical history review, physical examination (to assess changes in vision, balance, coordination and mental status), radiographic imaging (to determine the size, location, and consistency of tumor), histological evaluation (to detect the presence of atypical cells, the growth of new blood vessels, and the indicators of cell division called mitotic figures), immunohistochemical staining (to demonstrate binding for glial fibrillary acidic protein (GFAP), S100, and vimentin, etc), and molecular tests (to identify IDH1/2 mutations, p/19q codeletions, BRAF V600E and BRAF fusion mutations, ATRX, H3F3A, TERT, CIC and FUBP1 mutations, etc) [4].

2.6.1 Diffuse astrocytoma, IDH mutant (gemistocytic astrocytoma, IDH mutant)/diffuse astrocytoma, IDH wild-type/diffuse astrocytoma, NOS (WHO Grade II)

Diffuse astrocytoma usually occurs in the cerebrum (frontal and temporal lobes), brain stem, spinal cord, optic nerve, optic chiasm, optic pathway, hypothalamus, and thalamus. The tumor tends to distribute evenly throughout the white matter, resulting in a loss of the normal gray-white junction and making the central region ivory white. With a modest blood supply, the tumor appears less vascular than the adjacent brain. Radiographic imaging demonstrates asymmetry and/or enlargement of a region of the brain, along with abnormal density and signal. Histologic examination of unfrozen tissue reveals the increased cellular density, modest nuclear pleomorphism, increased intercellular edema, and occasional bubbly collections of fluid (so-called microcysts). Molecularly, diffuse astrocytoma, IDH mutant, harbors a mutated *IDH* gene (notably IDH-R132H); diffuse astrocytoma, IDH wild type, possesses an intact *IDH* gene; and diffuse astrocytoma, NOS, has undetermined *IDH* gene status. Gemistocytic astrocytoma, IDH mutant, is a variant of diffuse astrocytoma in which gemistocytes make up >20% of tumor cells and an IDH mutation is present [1].

2.6.2 Anaplastic astrocytoma, IDH mutant/anaplastic astrocytoma, IDH wild-type/anaplastic astrocytoma, NOS (WHO Grade III)

Anaplastic astrocytoma is a malignant tumor that represents an intermediate stage in the progression of diffuse astrocytoma to glioblastoma. Commonly found in the cerebrum, and occasionally in the cerebellum, brain stem, and spinal cord, anaplastic astrocytoma grows rapidly and often invades nearby healthy tissue. Macroscopically, the tumor may contain regions of lower grade diffuse astrocytoma (with an ivory white central region and modest blood supply) and higher grade glioblastoma (with a gray-purple region and rich blood supply). Histologically, the tumor displays frequent mitoses (which differs from diffuse astrocytoma), nuclear pleomorphism, and increased cellular density. Molecular determination of the status of the *IDH* and *TP53* genes, facilitates the classification of anaplastic astrocytoma, IDH mutant; anaplastic astrocytoma, IDH wild-type; and anaplastic astrocytoma, NOS [1].

2.6.3 Pilocytic astrocytoma (pilomyxoid astrocytoma) (WHO Grade I)

PA (also known as *juvenile pilocytic astrocytoma)* is a benign, slow-growing tumor that typically arises in the cerebellum, hypothalamus and third ventricular region in children and teens. Macroscopically, the tumor is a well-circumscribed mass with cystic portions filled with fluid and nodule. Radiographic imaging reveals a large cystic lesion with a brightly enhancing mural nodule and calcification (in 20% of cases). Histologically, the tumor shows bipolar cells with long, hair-like, GFAP-positive processes (so-called pilocytic), eosinophilic granular bodies, and microcysts. Molecularly, the tumor may exhibit gain in chromosome 7q34 involving the BRAF locus, activating mutations in *FGFR1* and *PTPN11*, as well as in *NTRK2* fusion genes.

PMA is a variant of PA, although PMA appears to behave more aggressively than PA. Histologically, PMA displays piloid and highly monomorphous cells in a uniform and extensive myxoid background, without Rosenthal fibers and eosinophilic granular bodies. In contrast, PA has biphasic architecture, with Rosenthal fibers and eosinophilic granular bodies. Further, PMA contains neoplastic cells (with the ability to infiltrate into the surrounding neural parenchyma), mitoses, and necrosis [1].

2.6.4 Subependymal giant cell astrocytoma (WHO Grade I)

Subependymal giant cell astrocytoma (SGCA) is a benign tumor (of >1 cm with calcification) that often occurs in the interventricular foramen (of Monro), the cerebral aqueduct (of Sylvius) and the third or fourth ventricle. Associated with tuberous sclerosis, SGCA typically affects patients of <20 years of age. Histologically, SGCA shows fibrillated spindle cells and globular large cells, with abundant eosinophilic cytoplasm; voluminous, eccentric nucleus; and large nucleoli, in addition to mitosis, necrosis, calcification, and perivascular lymphocytes. Its spindle cells stain positive for GFAP and S-100 protein. Molecularly, patients with tuberous sclerosis often harbor a mutation in one of two tuberous sclerosis genes (*TSC1*/hamartin or *TSC2*/tuberin), which results in an overexpression of the mammalian target of rapamycin (mTOR) complex 1, and increased risk of developing SGCA and subependymal nodules [1].

Differential diagnoses of astrocytomas include central neurocytoma, choroid plexus papilloma (CPP), choroid plexus carcinoma (CPC), anaplastic meningioma, ganglioglioma/glioneuronal tumors, oligodendroglioma, pleomorphic xanthoastrocytoma, primary cerebral and skull-based sarcomas, hypothalamic hamartoma, macrophage-rich lesions (including stroke and multiple sclerosis), metastatic malignancies, progressive multifocal leukoencephalopathy (PML), radiation necrosis, and subependymal nodule.

2.7 Treatment

The current treatment options for astrocytoma include surgery, observation, radiation therapy, chemotherapy, stem cell transplant, and targeted therapy. Whereas the treatment course for lower-grade astrocytoma (e.g., PA) is surgery alone, the standard care for high-grade astrocytoma consists of surgical resection, chemotherapy (e.g., vinblastine alone; temozolomide alone; carboplatin with or without vincristine; a combination of thioguanine, procarbazine, lomustine, and vincristine; or bevacizumab), and radiotherapy [4].

2.8 Prognosis

Low-grade astrocytoma (Grades I and II) portends a relatively favorable prognosis, although young patients with fibrillary histology, diencephalic syndrome, and intracranial hypertension at initial presentation have a poor prognosis. In patients with PA, the presence of a *BRAF–KIAA* fusion confers a better clinical outcome.

High-grade astrocytoma generally gives a poor prognosis in younger patients. Patients with anaplastic astrocytoma can survive for 5 years, while those undergoing gross-total resection or harboring IDH1/2 mutations (which are detected in 85% of lower-grade gliomas and only in 5% of glioblastomas) may fare better [4].

References

1. Louis DN, Ohgaki H, Wiestler OD. *WHO Classification of Tumours of the Central Nervous System*. 4th rev. ed. Lyon, France: IARC Press, 2016.
2. Appin CL, Brat DJ. Molecular genetics of gliomas. *Cancer J.* 2014;20(1):66–72.

3. Brandner S, von Deimling A. Diagnostic, prognostic and predictive relevance of molecular markers in gliomas. *Neuropathol Appl Neurobiol.* 2015;41(6):694–720.

4. PDQ Pediatric Treatment Editorial Board. *Childhood Astrocytomas Treatment (PDQ®): Health Professional Version.* PDQ Cancer Information Summaries. Bethesda, MD: National Cancer Institute (US), 2002.

3
Glioblastoma

Divya Khosla and Ritesh Kumar

3.1 Definition

Arising from glia or supporting tissue of the brain parenchyma, gliomas are generally categorized into Grade I (e.g., pilocytic astrocytoma and subependymal giant cell astrocytoma), Grade II (e.g., diffuse astrocytoma, fibrillary astrocytoma, ependymoma, pleomorphic xanthoastrocytoma, oligodendroglioma, mixed oligoastrocytoma, and optic nerve glioma), Grade III (e.g., anaplastic astrocytoma, anaplastic pleomorphic xanthoastrocytoma, anaplastic oligodendroglioma, and anaplastic oligoastrocytoma), and Grade IV (e.g., glioblastoma and diffuse midline glioma).

Glioblastoma (formerly *glioblastoma multiforme*), a WHO Grade IV tumor, represents the most common primary malignant brain tumor, with extremely aggressive behavior. Histologically, it is characterized by nuclear atypia, brisk mitotic activity, intense microvascular proliferation, and necrosis. Biologically, it shows extensive infiltration into the brain parenchyma, which makes complete resection impossible and recurrence almost a certainty.

The 2016 update of the WHO classification of central nervous system tumors incorporates both molecular parameters (particularly isocitrate dehydrogenase [IDH] status) and histopathologic features [1]. Molecularly, glioblastoma is a heterogeneous disease with many subtypes, including glioblastoma, IDH wild-type (consisting of giant cell glioblastoma, gliosarcoma, and epithelioid glioblastoma); glioblastoma, IDH mutant type; and glioblastoma not otherwise specified (NOS), in which full IDH evaluation cannot be performed [1,2].

Based on gene expression profiling, glioblastoma can be divided into four subtypes: classical, proneural, mesenchymal, and neural. The classical subtype contains chromosome 7 amplification, chromosome 10 deletion, epidermal growth factor receptor (EGFR) amplification, and homozygous deletion of the Ink4a/ARF locus. The mesenchymal subtype shows a high frequency of neurofibromatosis type 1 mutation/deletion and high expression of CHI3L1 and MET. The proneural subtype (including most secondary glioblastomas) is associated with younger age,

PDGFRA alterations, and IDH1 and TP53 mutations. The neural subtype is characterized by expression of neuronal markers [3].

3.2 Biology

Glioblastoma can arise *de novo* (i.e., primary glioblastoma) or result from malignant transformation of other gliomas, such as diffuse astrocytoma and/or anaplastic astrocytoma (i.e., secondary glioblastoma). As the deadliest neoplasm with the lowest survival rates among all brain tumors, glioblastoma notoriously infiltrates the adjacent brain parenchyma, due to the ability of tumor cells to travel long distances from the primary lesion and to extend across the midline to the contralateral cerebral hemisphere.

3.3 Epidemiology

Glioblastoma comprises 14.9% of all primary brain tumors and 46.6% of malignant brain tumors (46.6%). Glioblastoma has an annual incidence of 3.20 per 100,000 in the United States and appears to increase with age (the median age at diagnosis is 64 years, with the highest incidence in 75–84 years). Glioblastoma is 1.57 times more common in males and approximately 2 times higher among whites as compared to blacks [4].

Glioblastoma, IDH wild type, accounts for about 90% of cases and is analogous with the clinically defined primary or *de novo* glioblastoma seen in patients over 55 years of age, with a median age at diagnosis of 62 years. Glioblastoma, IDH mutant type, occurs in 10% of cases and is consistent with secondary glioblastoma with a prior history of diffuse astrocytoma or anaplastic astrocytoma. It is seen in comparatively younger patients with median age of diagnosis at 44 years.

3.4 Pathogenesis

Risk factors for the development of brain tumors include exposure to ionizing radiation (e.g., prior prophylactic cranial irradiation) [5], occupational and environmental exposures, and cumulative use of cellular and cordless telephones [6]. Genetic susceptibility and familial aggregation have also been linked with the development of brain tumors [7–8].

One of the initial steps in gliomagenesis is loss of cell cycle control due to mutations in tumor suppressor genes. Glioblastoma exhibits the highest degree of angiogenesis of all solid tumors. Another factor responsible for gliomagenesis is p53 mutations resulting in loss of cell cycle control and failure to undergo apoptosis. Various signaling pathways implicated in the pathogenesis of glioblastomas are RTK/PI3K/Akt, mTOR, the p53 pathway, ATM/Chk2/p53 pathway, Rb pathway, Ras/MAPK pathway, STAT3, and the glioma stem cell pathway.

Primary glioblastoma is characterized genetically by loss of heterozygosity (LOH) 10q (70% of cases), EGFR amplification (36%), p16INK4a deletion (31%), and PTEN mutations (25%). Secondary glioblastoma is characterized by *IDH1* mutations, which are associated with a hypermethylation phenotype. It is a definitive diagnostic molecular marker of secondary glioblastomas. TP53 mutation is the most common and earliest detectable genetic alteration, present in 60% of low-grade astrocytomas; LOH 10q (10q25-qter) is the most common genetic alteration observed in both primary and secondary glioblastomas [9].

3.5 Clinical features

Patients with glioblastoma often manifest with the classical triad of increased intracranial pressure (headache, nausea, and papilloedema). Other symptoms related to raised intracranial pressure or local pressure on sensitive intracranial structures are vomiting and drowsiness. Perilesional edema may cause more symptoms and neurological deficits (e.g., motor weakness, sensory loss, speech problems, or visual loss) than the tumor itself. Depending on the tumor location, seizures may be partial or generalized. Glioblastoma has a short clinical history of 3–6 months due to the rapidly progressive nature of the disease.

3.6 Diagnosis

Clinical history should be obtained from the patient and relatives. Complete neurological examination including higher mental functions, coordination, sensations, motor, reflexes, and cranial nerves should be performed. Fundus examination should be done to check papilloedema. Neuroradiologic imaging of choice in brain tumors is MRI, which is superior to CT in terms of demonstrating neuroanatomy and tumor volumes as well as image resolution. Histological confirmation of diagnosis is required for brain tumors and for appropriate management. Stereotactic biopsy is obtained for histological diagnosis if tumor is deeply situated, deemed unresectable, or is in an eloquent area or near eloquent area.

Glioblastoma is a large, poorly delineated tumor with an apparent pseu-docapsule and peritumoral edema. Its cut section is variegated, with a yellowish, necrotic central area, grayish peripheral rim, recent and old hemorrhages, and cysts due to liquefied necrotic tumor tissue.

On MRI, glioblastoma usually appears as an irregularly shaped hypodense lesion with a peripheral ring-like zone of contrast enhancement and sur-rounding edema. It demonstrates low signal intensity on T1-weighted images and high signal intensity on T2-weighted images. The central hypointense core corresponds to the area of necrosis.

Characteristic histopathological features of glioblastoma include nuclear atypia, mitotic activity, microvascular proliferation, and necrosis. The tumor cells are polygonal to spindle-shaped with indistinct cell borders, increased nuclear-to-cytoplasmic ratio, and nuclear pleomorphism. The tumor stains positive for glial fibrillary acidic protein, vimentin, S-100, and AE1–AE3 (>95%) but negative for CAM5.2, CK7, CK20, and BerEP4.

At the molecular level, glioblastoma is associated with amplification of EGFR (particularly in the small cell variant); mutations of p16, p53, and *PTEN*; and loss of heterozygosity at 10q. Interestingly, primary glioblastoma (which arises *de novo*, without recognizable precursor lesions) often contains a p53 mutation, whereas secondary glioblastoma (which develops slowly from Grade II or III astrocytoma) tends to harbor partial *PTEN* deletion.

3.7 Treatment

The current standard care for glioblastoma consists of safe maximal surgi-cal resection followed by concurrent radiation (60 Gy in 30 fractions over 6 weeks plus temozolomide 75 mg/m², 7 days per week for 6 weeks from the first to the last day of radiotherapy) followed by adjuvant temozolomide (150–200 mg/m² for 5 days every 28 days) [10].

Surgery helps establish diagnosis as it allows sampling of sufficient tissue material for histopathological and molecular analyses, as well as symptom-atic relief from mass effect. In general, gross tumor resection provides a significant survival benefit and decreased recurrence in comparison with subtotal tumor resection.

Whole brain radiotherapy followed by cone-down boost has proven valu-able for treating glioblastoma. The major guidelines for target delineation in glioblastoma were established by the European Organisation for Research and Treatment of Cancer (which calls for a single-phase technique) and the

Radiation Therapy Oncology Group (which calls for a two-phase cone-down technique).

Glioblastoma is characterized by inherent resistance to chemotherapy drugs. Temozolomide is an alkylating agent that represents a first-line treatment for glioblastoma and a second-line treatment for astrocytoma. *Pneumocystis carinii* pneumonia prophylaxis is required in patients receiving concurrent radiation and temozolomide.

Glioblastoma is the most angiogenic and VEGF has emerged as an important molecular target. Bevacizumab (humanized monoclonal antibody against VEGF) does not improve overall survival (OS) but seems to prolong median progression free survival (PFS). In newly diagnosed glioblastoma patients with nonmethylated MGMT promoter, bevacizumab plus irinotecan resulted in a superior PFS at 6 months and median PFS compared with temozolomide, but it did not improve OS.

3.8 Prognosis

The prognosis of glioblastoma remains poor despite advancements in treatment. Younger age at diagnosis and good performance status are independent favorable prognostic factors.

Recursive partitioning analysis categorizes patients into different risk groups based on tumor size and location, age at diagnosis, and KPS at presentation and treatment. The lowest risk group includes patients <40 years with tumor in the frontal lobe only. The intermediate-risk group comprises patients aged 40–65 years, KPS more than 70, and subtotal or total resection. The highest risk group includes patients >65 years, between 40 and 65 years with KPS less than 80, or biopsy only. Methylation of O6-methylguanine-DNA methyltransferase (MGMT), which is present in approximately half of glioblastoma patients over the age of 70, is associated with improved outcome in comparison to unmethylated MGMT and is considered a marker of better therapeutic response. Patients with glioblastoma (especially of the proneural subtype) containing CpG island methylator phenotype and IDH1 mutation as well as temozolomide-based chemoradiation have a significant OS advantage.

Glioblastoma exhibits heterogeneity at the morphological, biological, genomic, and antigenic levels, rendering the tumor cells resistant to available treatment modalities. The survival rates are bleak with a median survival of 3 months for untreated cases. Despite aggressive treatment, median survival is only 14.6 months with a 2-year OS of 27% and a 5-year survival rate of <10%. Glioblastoma eventually recurs despite treatment. Patients who progress >18 months after their initial treatment have significantly greater 2- and 5-year postprogression survival and OS. The conditional probability of survival into the future is favorable for patients surviving past 2 years from diagnosis as compared to newly diagnosed patients.

References

1. Louis DN, Perry A, Reifenberger G, et al. The 2016 World Health Organization classification of tumors of the central nervous system: A summary. *Acta Neuropathol.* 2016;131:803–20.
2. Cohen A, Holmen S, Colman H. IDH1 and IDH2 mutations in gliomas. *Curr Neurol Neurosci Rep.* 2013;13:345.
3. Verhaak RG, Hoadley KA, Purdom E, et al. An integrated genomic analysis identifies clinically relevant subtypes of glioblastoma characterized by abnormalities in PDGFRA, IDH1, EGFR and NF1. *Cancer Cell.* 2010;17:98.
4. Ostrom QT, Gittleman H, Xu J, et al. CBTRUS statistical report: Primary brain and central nervous system tumors diagnosed in the United States in 2009–2013. *Neuro Oncol.* 2016;18 Suppl 5:v1–75.
5. Neglia JP, Meadows AT, Robison LL, et al. Second neoplasms after acute lymphoblastic leukemia in childhood. *N Engl J Med.* 1991;325:1330–6.
6. Hardell L, Carlberg M, Hansson Mild K. Pooled analysis of two case-control studies on use of cellular and cordless telephones and the risk for malignant brain tumours diagnosed in 1997–2003. *Int Arch Occup Environ Health.* 2006;79:630–9.
7. Kyritsis AP, Bondy ML, Xiao M, et al. Germline p53 gene mutations in subsets of glioma patients. *J Natl Cancer Inst.* 1994;86:344–9.
8. Kyritsis AP, Bondy ML, Rao JS, Sioka C. Inherited predisposition to glioma. *Neuro-Oncol.* 2010;12:104–13.
9. Furnari FB, Fenton T, Bachoo RM, et al. Malignant astrocytic glioma: Genetics, biology, and paths to treatment. *Genes Dev.* 2007;21:2683–710.
10. Khosla D. Concurrent therapy to enhance radiotherapeutic outcomes in glioblastoma. *Ann Transl Med.* 2016;4:54.

4

Pleomorphic Xanthoastrocytoma

Xintong Zhao and Guoquan Jiang

4.1 Definition

Pleomorphic xanthoastrocytoma (PXA) is a rare primary brain tumor occurring in children and young adults, who typically present with temporal lobe epilepsy. As a WHO Grade II tumor with a cystic component and vivid contrast enhancement, PXA demonstrates several characteristic features: an astrocytic tumor unique in its superficial cortical location, histological variability (pleomorphism), prominent xanthomatous cells, distinct clinical course, and favorable response to surgical resection.

A variant of PXA with significant mitotic activity (5+ mitoses per 10 high-power field [HPF]) or with areas of necrosis is recognized as PXA with anaplastic features or anaplastic pleomorphic xanthoastrocytoma (APXA, WHO Grade III) by the 2016 *WHO Classification of Tumours of the Central Nervous System*. APXA may be either primary or secondary, although secondary APXA is more frequently reported than primary APXA. In addition, the recurrence interval in secondary APXA is longer than that in primary APXA, and secondary APXA seems to have a worse prognosis than primary APXA.

4.2 Biology

Literature regarding the biology of PXA is scarce and incomplete. However, a few studies using immunological, cytological, and molecular approaches have made significant contributions to the present understanding of PXA [1]. Firstly, PXA is genotypically distinct from other WHO Grade II gliomas in pathological process (see Section 4.4). Secondly, there are two unique features of PXA. One relates to the tendency of these tumors to exhibit a dichotomous astrocytic/neuronal genotype and phenotype. Glial fibrillary acidic protein (GFAP) and S-100 protein are consistently present, suggesting a primarily astrocytic phenotype. The other unique feature of PXA relates to the neuronal differentiation on account of reported neuronal markers, including Class III beta tubulin, synaptophysin, neurofilament proteins, SMI-31, and MAP2. Third, ultrastructural analyses support this

Table 4.1   Key Characteristics of PXA

Primarily Astrocytic Phenotype	GFAP and S-100 Protein
Neuronal markers	Class III beta tubulin, synaptophysin, neurofilament proteins, SMI-31, and MAP2 immunopositivity
Ultrastructural analyses	Astrocytic (intermediate filaments, lipid droplets, lysosomes) and neuronal (microtubules, dense core granules) differentiation

Note: GFAP, glial fibrillary acidic protein; PXA, pleomorphic xanthoastrocytoma.

dichotomy, demonstrating features consistent with both astrocytic (intermediate filaments, lipid droplets, lysosomes) and neuronal (microtubules, dense core granules) differentiation (Table 4.1).

4.3 Epidemiology

PXA is a rare tumor of the central nervous system. It was first reported as a unique histopathological entity in 1979, accounting for less than 1% of brain tumors [2]. The majority of PXA patients are children and teenagers (with an average age at diagnosis of 12 years), and the most common tumor location is the temporal lobe.

4.4 Pathogenesis

The pathogenesis of PXA remains inadequately elucidated. Up to the present, heredity abnormalities have mainly been studied. According to the existing literature, chromosomal losses and gains, translocation of chromosomes, mutation of genes, and other causes are involved (Table 4.2).

Table 4.2   Genetic Abnormalities of PXA

Losses of chromosome	The most common region: 9p21.3
	The less common regions: 4qter, 6qter, 8p, 10p, 13, 17pter, 18qter, 21qter, chromosomes 20 and 22
Gains of chromosomes	Almost all the chromosomes except for 13, 18, 22, and Y
Translocation	chromosome 1
Genetic mutation	The hot spot: BRAF V600E
	Others: CD34, MDM2
	Infrequent: p53, IDH1, IDH2
Others	Diploid or polyploid karyotypes
	Telomeric breakage and fusion events: 1, 15, 20, and 22

- *Chromosomal losses*: DNA loss on chromosome 9 appears to be the most common regional chromosomal abnormality in PXA, especially 9p21.3. Other less common regional losses involve chromosome 22, as well as 4qter, 6qter, 8p, 10p, 13, 17pter, 18qter, 21qter, and chromosome 20.
- *Gains of chromosomes*: Almost every chromosome except for chromosomes 13, 18, 22, and Y.
- *Translocation of chromosomes*: Translocations have been described involving chromosome 1.
- *Genetic mutation*: Many molecular features that are commonly observed in other WHO Grade II gliomas are generally absent in PXA [3]. TP53, IDH1, and IDH2 mutations are uncommon. Conversely, *BRAF* mutations (particularly at the V600E "hot spot") are common in PXA [4]. PXA is unique in its expression of wild-type or truncated CD34 as well as in its expression of MDM2 without gene amplification.
- *Other causes*: Diploid or polyploid karyotypes; telomeric breakage and fusion events have been reported for chromosomes 1, 15, 20, and 22.

4.5 Clinical signs

Supratentorial PXA is the most common, although the lesion is also described in other locations, such as cerebellum, spine, and pineal body. Spinal PXA is extremely rare. The clinical manifestations of PXA vary from no symptoms to headaches, dizziness, and sudden seizures.

4.6 Diagnosis

Typical radiologic findings of supratentorial lesions are cystic neoplasm accompanied by mural nodules. On CT scan, PXA appears isodense to hypodense. Calcification is uncommon. The cystic mass is hypointense on T1-weighted imaging and hyperintense on T2-weighted imaging [5]. Mural nodules are strongly enhanced after the administration of gadolinium. Peritumor edema is minimal (Figure 4.1).

Histologically, PXA shows a relatively solid growth pattern, composed of a combination of spindle cells, multinucleated giant cells, and foamy lipid-laden xanthomatous cells, in association with both pale and bright eosinophilic granular bodies (Figure 4.2). Other features include perivascular lymphocytic cuffing, scattered eosinophilic granular bodies, and a reticulin-rich network. The underlying cortex contains an infiltrative astrocytic component. Hemorrhage and protein granular degeneration

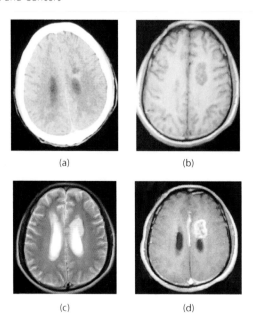

(a)   (b)

(c)   (d)

Figure 4.1   Cerebral radiological features: (a) periventricular hypodense lesion on CT scan; (b) hypointense signal on T1-weighted imaging; (c) hyperintense signal on T2-weighted imaging; and (d) enhanced MRI scan showing a heterogeneous and annular enhancement.

are variably present. Mitosis and necrosis are commonly absent except in tumors with anaplastic features (Figure 4.2). PXA is immunohistochemically positive for GFAP, S-100, reticulin, and class III beta tubulin (73%) but negative for chromogranin and p53 (or focal).

APXA often displays high mitotic activity (between 5 and 15 per 10 HPF) and Ki-67 proliferation index (20%–80%), vascular endothelial proliferation, intranuclear inclusions, xanthomatous cells, monotonous rhabdoid-like cells, perivascular lymphocytes, desmoplasia, reticulin fibers, calcification, and necrosis. APXA may be focally or weakly positive for GFAP, as well as positive for synaptophysin, Class III beta tubulin, and neurofilament protein.

Differential diagnoses include other cortical tumors—for example, ganglioglioma (less prominent contrast enhancement, calcification in ~50% of cases, no dural tail sign), dysembryoplastic neuroepithelial tumors (bubbly appearance, rare contrast enhancement), oligodendroglioma (frequent calcification), desmoplastic infantile ganglioglioma (young children, prominent dural involvement, multiple large lesions), and cystic meningioma—glioblastoma, giant cell glioblastoma, and malignant fibrous histiocytoma.

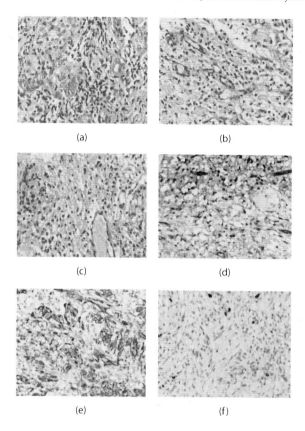

Figure 4.2   Characteristic histological features of pleomorphic xanthoastrocytoma: (a) photomicrograph of the lesion showing giant and spindle-shaped tumor cells. The cells include mono- or multinucleated giant astrocytes with a foamy or vacuolated cytoplasm (hematoxylin and eosin, or H&E; original magnification, 400×); (b) pale and intensively eosinophilic granular bodies with occasional xanthic tumor cells (H&E, 400×); (c) the tumor cells are separated by a reticular fiber network (H&E, 400×); (d) immunohistochemical examination showing S-100 positivity (original magnification, 400×); (e) immunohistochemical examination showing GFAP positivity (original magnification, 400×); and (f) approximately 2% of tumor cells are positive for Ki-67 (original magnification, 200×).

4.7 Treatment

Most reports suggest that surgery should be the first choice. Patients can have a favorable outcome after gross total resection [6,7]. However, the best choice of treatment depends on many individual factors, including the patient's medical history and overall health condition; the type, location, and size of the tumor; and how slowly or quickly the tumor is expected to progress. If surgery is performed and the tumor is completely resected, further treatment may not be required. The patient will, however, need repeated MRI to monitor the area for tumor regrowth. For tumors that recur, another surgical resection might be attempted. For tumors that cannot be completely removed, adjuvant radiation and chemotherapy may also be recommended. However, their effects remain unclear.

4.8 Prognosis

PXA is a rare glial neoplasm, originating from subpial astrocytes, that has relatively favorable outcomes due to its WHO Grade II designation. The overall survival rate at 5 years is 74% [8]. However, up to 20% of PXA cases may undergo malignant transformation to APXA or glioblastoma, while some others may suffer recurrence or dissemination by meninges or cerebrospinal fluid. Nonetheless, compared to glioblastoma, even high-grade PXA patients have significantly better overall rates of survival [9,10].

References

1. Ida CM, Rodriguez FJ, Burger PC, et al. Pleomorphic xanthoastrocytoma: Natural history and long-term follow-up. *Brain Pathol.* 2015;25:575–86.
2. Giannini CPW, Louis DN, Liberski P. Pleomorphic xanthoastrocytoma. In: Louis DN, Ohgaki H, Wiestler OD, Cavanee WK, eds. *WHO classification of tumors of the central nervous system.* Lyon: IARCP, 2007;22–24.
3. Martinez R, Carmona FJ, Vizoso M, et al. DNA methylation alterations in grade II- and anaplastic pleomorphic xanthoastrocytoma. *BMC Cancer.* 2014;14:213.
4. Lohkamp LN, Schinz M, Gehlhaar C, et al. MGMT Promoter methylation and BRAF V600E mutations are helpful markers to discriminate pleomorphic xanthoastrocytoma from giant cell glioblastoma. *PLoS One.* 2016;11:e0156422.
5. Moore W, Mathis D, Gargan L, et al. Pleomorphic xanthoastrocytoma of childhood: MR imaging and diffusion MR imaging features. *Am J Neuroradiol.* 2014;35:2192–6.
6. Dodgshun AJ, Sexton-Oates A, Saffery R, et al. Pediatric pleomorphic xanthoastrocytoma treated with surgical resection alone: Clinicopathologic features. *J Pediatr Hematol Oncol.* 2016; 38(7):e202–6.

7. Gallo P, Cecchi PC, Locatelli F, et al. Pleomorphic xanthoastrocytoma: Long-term results of surgical treatment and analysis of prognostic factors. *Brit J Neurosurg.* 2013;27:759–64.

8. Gaba P, Puffer RC, Hoover JM, et al. Perioperative outcomes in intracranial pleomorphic xanthoastrocytoma. *Neurosurgery.* 2016 May 12. [Epub ahead of print]. DOI: 10.1227/NEU.0000000000001250 (Epub).

9. Sharma A, Nand Sharma D, Kumar Julka P, et al. Pleomorphic xanthoastrocytoma—A clinico-pathological review. *Neurol Neurochirurg Pol.* 2011;45:379–86.

10. Perkins SM, Mitra N, Fei W, et al. Patterns of care and outcomes of patients with pleomorphic xanthoastrocytoma: A SEER analysis. *J Neuro-Oncol.* 2012;110:99–104.

5
Chordoid Glioma, Angiocentric Glioma, and Diffuse Midline Glioma

5.1 Definition

Chordoid glioma (also known as *chordoid glioma of the third ventricle*) is a rare, noninvasive, slow-growing brain tumor. Mostly originating from the anterior part of the third ventricle and occasionally affecting the temporoparietal region, left thalamus, and the corona radiata/thalamus in children, chordoid glioma (WHO Grade II) is noted for the histological presence of both glial and chordoid elements, reminiscent of chordoma, and the avid staining with glial fibrillary acidic protein (GFAP).

Angiocentric glioma is a slow-growing supratentorial tumor characterized by its striking perivascular pattern of growth and "dot-like" epithelial membrane antigen (EMA) staining of microlumens. With ependymomatous differentiation of lesion cells, angiocentric glioma (WHO Grade I) appears radiographically similar to other low-grade astrocytomas.

Diffuse midline glioma is a common type of diffuse intrinsic pontine glioma (DIPG) that arises within the pons (called *pontine glioma* or *diffuse intrinsic brainstem glioma*). Pontine glioma is usually high grade and locally infiltrative, displaying histological similarity to anaplastic astrocytoma (WHO Grade III) or glioblastoma (WHO Grade IV) and carrying a specific point mutation (K27M) in histone H3 (called *diffuse midline glioma H3 K27M–mutant*, WHO Grade IV). In contrast, some 20% of brainstem tumors affect the cervicomedullary junction and tectum (known as *nonpontine glioma*). Nonpontine glioma is usually a low-grade, discrete, and well-circumscribed astrocytoma (e.g., pilocytic astrocytoma WHO Grade I), although about 10%–20% of nonpontine gliomas may be high grade and behave similarly to DIPGs [1,2].

5.2 Biology

Chordoid glioma likely derives from ependymal cells or tanycytes (which are primitive progenitor cells of ependymal cells and glial cells, predominantly present in the anterior portion of the third ventricle), as chordoid glioma tumor cells often demonstrate features of ependymal differentiation on electron microscopy, including basal lamina, microvilli, and intermediate filaments. Typically occurring in the anterior portion of the third ventricle (known as *chordoid glioma of the third ventricle*) or the suprasellar region, chordoid glioma often extends into the hypothalamus and occasionally into the juxtaventricular white matter and thalamus.

Angiocentric glioma is postulated to originate from astrocytic and ependymal lineages, or radial glia or neuronal lineages. As a supratentorial tumor, angiocentric glioma often affects the temporal lobe (epileptogenic foci).

Diffuse midline glioma arises from glial cells located in the pons, thalamus, and spinal cord, as well as the third ventricle, hypothalamus, pineal region, and cerebellum.

5.3 Epidemiology

Chordoid glioma is a rare, low-grade neuroepithelial tumor (with <100 cases described in the English literature) that mainly affects adults (mean age of 45 years at diagnosis) and shows a female predominance (female-to-male ratio of 2 to 1).

Angiocentric glioma is a rare, low-grade tumor with <100 cases reported to date. It is associated with seizures in children (mean age of 6.5 years at diagnosis, range of 2–14 years).

Diffuse midline glioma accounts for 80% of brainstem tumors, of which about 300 pediatric cases and 100 adult cases are reported each year in the United States. Diffuse midline glioma located in the pons often affects children (median age of 5–9 years at diagnosis), and that located in thalamus and spinal cord involves youth (median age of 24–25 years at diagnosis) [2].

5.4 Pathogenesis

Chordoid glioma is a homogeneous tumor of uncertain histogenesis with distinct clinicopathologic features. Similarly, the cytogenetic basis of angiocentric glioma remains unclear.

Diffuse midline glioma (and DIPG) is linked to chromosomal and genomic abnormalities, including (i) somatic mutation in the histone H3.1 (*H3F3A*) or H3.3 (*HIST1H3B*) genes encoding the histone H3 variants, H3.1 and H3.3; (ii) activating mutation in the activin A receptor, Type I (*ACVR1*) gene, which may cause the autosomal dominant syndrome fibrodysplasia ossificans progressiva (this mutation often occurs concurrently with H3.3 mutations); (iii) receptor tyrosine kinase (*PDGFRA*) amplification; and (iv) *TP53* gene deletion on chromosome 17p. Interestingly, in diffuse midline glioma (and DIPG), histone H3-K27M mutation is mutually exclusive with IDH1 mutation and epidermal growth factor receptor (EGFR) amplification, as well as *BRAF-V600E* mutation. H3K27 alterations appear to be the founding event in DIPG and the mutations in the two main histone H3 variants drive two distinct oncogenic programs with potential specific therapeutic targets [1,2].

5.5 Clinical features

The clinical symptoms of chordoid glioma are mostly attributable to tumor-related intracranial hypertension; obstructive hydrocephalus; and compression of hypophysis (pituitary gland), optic chiasm, and hypothalamus, leading to headache, nausea, vomiting, ataxia, visual impairment, lethargy, somnolence, urinary incontinence, hypothyroidism, diabetes insipidus, amenorrhea, weight loss or gain, memory deficits, confusion, bitemporal hemianopsia, seizures, syncopal episodes, and speech difficulties [3].

Angiocentric glioma often presents with intractable seizures, headache, decreasing visual acuity, blank stares, episodes of stomach sensation, and speech arrest [4].

Clinically, diffuse midline glioma may manifest in headaches, nausea, diplopia, lethargy (related to increased intracranial mass), seizures (related to brain irritation), hemiparesis and dysphasia (related to brain invasion), and focal neurologic deficits.

5.6 Diagnosis

Chordoid glioma of the third ventricle is a solid, well-circumscribed mass (mean diameter of 3.1 cm, range 1.5–7.0 cm) with occasional formation of intratumoral calcification or cysts. On CT, the tumor often appears as an ovoid, hyperdense lesion; on MRI, it displays uniform contrast enhancement on T1-weighted images and slight hypersignal intensity on T2-weighted images. Histologically, chordoid glioma shows cords and clusters of epithelial cells with eosinophilic cytoplasm and relatively uniform, round to oval nuclei embedded in mucinous stroma. Immunohistochemically, chordoid glioma is diffusely positive for GFAP, vimentin, and CD34 and focally positive for EMA and cytokeratin [3,5,6]. Other tumors originating in the anterior portion of the third ventricle include ependymomas, central neuro-cytomas, craniopharyngiomas, and suprasellar meningiomas. As these neo-plasms are mostly negative for CD34, use of CD34 may help differentiate them from chordoid glioma [6].

Angiocentric glioma shows a stable, cyst-like lesion that is focal, nonen-hancing on MRI (T1 hypointense, T2 hyperintense). Histologically, the tumor is defined by diffusely infiltrating monomorphous bipolar spindled cells arranged in angiocentric pattern about cortical blood vessels; epen-dymoma-like pseudorosettes, subpial palisading, miniature schwannoma-like nodules; and rare EMA-positive cytoplasmic dots (suggestive of partial ependymal differentiation). Most angiocentric gliomas are positive for GFAP and contain characteristic EMA-positive microlumens (suggestive of epen-dymal lesion) [4]. Angiocentric glioma should be considered in the neo-plastic differential diagnosis of medically refractory epilepsy in children and young adults [7].

Diffuse midline glioma is often diagnosed clinically on the basis of neurolog-ical signs, duration of symptoms, and specific neuro-imaging findings [2]. On CT, diffuse midline glioma is typically hypodense with little enhancement. On MRI, diffuse midline glioma shows decreased intensity on T1, appears heterogeneously increased on T2, usually minimal on T1 C+ (Gd), and usually normal, occasionally mildly restricted on DWI/ADC. Histologically, diffuse midline glioma may show a wide spectrum of features, including gliomas with giant cells, epithelioid and rhabdoid cells, primitive neuroectodermal tumor–like foci, neuropil-like islands, pilomyxoid features, ependymal-like areas, sarcomatous transformation, ganglionic differentiation, and pleo-morphic xanthoastrocytoma–like areas. Immunohistochemically, diffuse midline glioma is positive for S100, NCAM1, OLIG2, p53 protein (50%), and GFAP (variable); but negative for chromogranin-A and NeuN. Molecularly, diffuse midline glioma often contains a histone H3-K27M mutation, which is mutually exclusive with *IDH1* mutation and *EGFR* amplification and which rarely co-occurs with *BRAF-V600E* mutation. The tumor is commonly asso-ciated with p53 overexpression, ATRX loss (except in pontine gliomas), and monosomy 10 [1,2].

5.7 Treatment

Surgical resection represents a preferred option for treating chordoid glioma, whereas radiotherapy (10.5–12 Gy) is reserved as adjuvant therapy for the management of residual tumor following resection [8]. Postoperative com-plications for chordoid glioma include hypothalamic dysfunction, diabetes insipidus, syndrome of inappropriate antidiuretic hormone, panhypopituitarism, weight gain, short-term memory deficits, severe amnesia, hematoma formation, bacterial meningitis, and pulmonary embolism [8].

Angiocentric glioma is amenable to neurosurgical intervention, and the outcome following gross total resection (GTR) of the tumor is excellent without need for radiation or chemotherapy. However, for patients undergoing subtotal tumor resection, postoperative radiotherapy facilitates effective control of seizures.

Diffuse midline glioma (and DIPG) cannot be surgically removed, due to its location and the infiltrative nature of the disease. Use of radiotherapy or chemotherapy may be considered.

5.8 Prognosis

Chordoid glioma is a low-grade tumor with a poor prognosis due to its location and the difficulty in performing complete surgical resection without causing severe hypothalamic damage. Partial resection of the tumor is often associated with high recurrence rates [3,9].

Angiocentric glioma is an indolent brain tumor with an excellent outcome and postoperative seizure freedom rate of 100% following GTR [10].

In cases when GTR is not feasible, subtotal resection together with postoperative radiotherapy and/or chemotherapy enhances seizure control and progression-free survival.

Diffuse midline glioma (and DIPG) has a poor prognosis, with a median survival of <1 year and <10% of children surviving for >2 years.

References

1. Castel D, Philippe C, Calmon R, et al. Histone H3F3A and HIST1H3B K27M mutations define two subgroups of diffuse intrinsic pontine gliomas with different prognosis and phenotypes. *Acta Neuropathol.* 2015;130(6):815–27.
2. Solomon DA, Wood MD, Tihan T, et al. Diffuse midline gliomas with histone H3-K27M mutation: A series of 47 cases assessing the spectrum of morphologic variation and associated genetic alterations. *Brain Pathol.* 2016;26(5):569–80.
3. Destefani MH, Mello AS, de Oliveira RS, Simão GN. Chordoid glioma of the third ventricle. *Radiol Bras.* 2015;48(5):338–9.

4. Lum DJ, Halliday W, Watson M, Smith A, Law A. Cortical ependymoma or monomorphous angiocentric glioma. *Neuropathology.* 2008;28:81–6.

5. Bongetta D, Risso A, Morbini P, Butti G, Gaetani P. Chordoid glioma: A rare radiologically, histologically, and clinically mystifying lesion. *World J Surg Oncol.* 2015;13:188.

6. Ki SY, Kim SK, Heo TW, Baek BH, Kim HS, Yoon W. Chordoid glioma with intraventricular dissemination: A case report with perfusion MR imaging features. *Korean J Radiol.* 2016;17(1):142–6.

7. Preusser M, Hoischen A, Novak K, et al. Angiocentric glioma report of clinico-pathologic and genetic findings in 8 cases. *Am J Surg Pathol.* 2007;31:1709–18.

8. PDQ Pediatric Treatment Editorial Board. *Childhood Brain Stem Glioma Treatment (PDQ®): Health Professional Version.* PDQ Cancer Information Summaries. Bethesda, MD: National Cancer Institute (US), 2002.

9. Ampie L, Choy W, Lamano JB, et al. Prognostic factors for recurrence and complications in the surgical management of primary chordoid gliomas: A systematic review of literature. *Clin Neurol Neurosurg.* 2015;138:129–36.

10. Shakur SF, McGirt MJ, Johnson MW, et al. Angiocentric glioma: a case series. *J Neurosurg Pediatr.* 2009;3(3):197–202.

6

Astroblastoma

Mohammed Hmoud and Alaa Samkari

6.1 Definition

Astroblastoma is a rare glial tumor of uncertain origin. It is classified as either low or high grade (well differentiated or anaplastic/malignant). This classification is based on the cellularity, presence of necrosis, and mitotic figures [1]. Astroblastoma constitutes up to 3% of all neuroglial tumors such as astrocytomas, oligodendrogliomas, glioblastomas, and others.

6.2 Biology

Although it was described by Bailey and Cushing and further characterized by Bucy and Cushing in the 1920s, astroblastoma has no definite biological origin. The earliest theory developed suggests that astroblasts are the precursors for this type of tumors. Other theories state that the cellular origin of astroblastoma arises from dedifferentiation from mature astroglial cells; yet another theory suggests that these cells are intermediates between astrocytes and ependymal cells.

6.3 Epidemiology

Astroblastoma commonly occurs in children (aged 5–10 years) and adults (aged 21–30 years). It rarely presents as a congenital mass or among the elderly. A recent review revealed a mean age of 7.7 years (range 30 days to 60 years) [1]. Astroblastoma appears to have a female gender predilection, and a recent review found that 72.6% of cases involved are females [1–3].

6.4 Pathogenesis

Genetic studies were to some degree helpful in obtaining insights about astroblastoma pathogenesis and histological origin. Reported findings were gains in chromosomes 20q and 19, monosomies of chromosomes 10, 21, and 22, and loss of heterogeneity in chromosome 9s and 19 [4].

6.5 Clinical features

Astroblastoma presents as any mass-occupying lesion. In a recent review, headache, nausea, vomiting, and seizure were described as the most common manifestations [1]. In contrast, cognitive and behavioral changes are rare. Magnetic resonance imaging demonstrates that astroblastoma usually appears as supratentorial, superficial, well-defined, cystic, solid enhancing lesions affecting the frontal lobe followed by parietal and temporal lobes.

6.6 Diagnosis

Diagnosis of astroblastoma can be challenging, because it is rare and shares common features with other glial tumors, especially astrocytoma and ependymoma.

Macroscopically, astroblastoma is a large, peripherally located, supratentorial, lobulated, solid, cystic mass with little associated vasogenic edema and a bubbly appearance due to the presence of multiple cysts. Calcification in a punctate pattern is seen in 85% of cases. On MRI, the tumor appears iso- to hypointense on T1 and heterogeneously hyperintense on T2/FLAIR, with heterogeneous enhancement on T1 Gd (C+).

Histologically, astroblastoma is characterized by short processes forming perivascular pseudorosettes, with thick processes from the cell body to the adventitia of the vessel, and vascular hyalinization with little fibrillar background (Figure 6.1). Hypercellularity, high mitotic index, and the presence of vascular proliferation or necrosis with pseudopalisading, as well as occasional signet-ring cells, are suggestive of anaplastic astroblastoma. Immunohistochemically, astroblastoma shows different degrees of positivity to S-100, vimentin, and glial fibrillary acidic protein and is negative for phosphotungstic acid hematoxylin (PTAH), synaptophysin, and cytokeratin [3].

Differential diagnoses include glial tumors (e.g., pilocytic astrocytoma, pleomorphic xanthoastrocytoma, ependymoma, and oligodendroglioma) and other nonglial tumors (e.g., meningioma, schwannoma, and lymphoma), as well as metastasis from cancer in the lung, breast, colon, or melanoma.

Note that pilocytic astrocytoma contains biphasic piloid areas with Rosenthal fibers alternating with spongy microcystic areas with eosinophilic granular bodies. Pleomorphic xanthoastrocytoma shows a fascicular pattern, pleomorphic cells, lipidized cells, eosinophilic granular bodies, and perivascular lymphocytes; oligodendroglioma has nodular calcifications

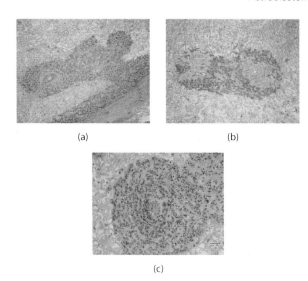

(a)   (b)

(c)

Figure 6.1   Routine hematoxylin and eosin histology, showing a well-differentiated astroblastoma: (a, b) low-power magnification demonstrating astroblastic pseudorosette; (c) medium-power magnification demonstrating astroblastic pseudorosette, perivascular hyalinization, and abundant elongated eosinophilic cells.

instead of punctate calcifications. Ependymoma is more fibrillar and shows smaller and less pleomorphic nuclei, true rosettes, and less sclerosis [5].

6.7 Treatment

The mainstay of treatment of astroblastoma is surgical intervention because most tumors are well circumscribed and not infiltrative. The majority of patients reported in the literature underwent gross total resection. In high-grade tumors, radiation and chemotherapy may be needed. In a recent review, chemotherapy was used in 18% of the reviewed cases, whereas radiation was needed in 51% [1]. Chemotherapeutic agents used in the literature include temozolomide, vincristine, carboplatin, cisplatin, and etoposide. The therapeutic radiation dose reported in the literature ranged from 36 to 72 Gy.

6.8 Prognosis

Although the progression for astroblastoma is unpredictable, low-grade tumors tend to be less infiltrative with far fewer symptoms and complications. Almost 27% of patients go through recurrence regardless of their tumor grade. Anaplastic astroblastoma predicts poor prognosis.

References

1. Samkari A, Hmoud M, Al-Mehdar A, Abdullah S. Well-differentiated and anaplastic astroblastoma in the same patient: A case report and review of the literature. *Clin Neuropathol.* 2015;34(6):350.

2. Cunningham DA, Lowe LH, Shao L, Acosta NR. Neuroradiologic characteristics of astroblastoma and systematic review of the literature: 2 new cases and 125 cases reported in 59 publications. *Pediatr Radiol.* 2016;46(9):1301–8.

3. Narayan S, Kapoor A, Singhal MK, Jakhar SL, Bagri PK, Rajput PS, Kumar HS. Astroblastoma of cerebrum: A rare case report and review of literature. *J Cancer Res Therapeut.* 2015;11(3):667.

4. Hata N, Shono T, Yoshimoto K, et al. An astroblastoma case associated with loss of heterozygosity on chromosome 9p. *J Neuro-Oncol.* 2006;80(1):69–73.

5. Bailey P, Cushing HA. *A Classification of Tumors of the Glioma Group on a Histogenetic Basis with a Correlated Study of Prognosis.* JB Lippincott Co. Philadelphia, PA. 1926, p. 175.

7
Gliomatosis Cerebri

Vamsi Krishna Yerramneni and Omekareswar Rambarki

7.1 Definition

The term *gliomatosis cerebri* (GC) was first coined by Nevinin in 1938, who described three cases with a diffuse overgrowth of neuroglial cells in wide areas of the cerebral hemisphere [1,2]. The 2007 WHO classification of brain tumors defined *GC* as a diffuse glioma with an extensive infiltration of a large region of the central nervous system (CNS) and the involvement of at least three contiguous cerebral lobes. However, GC was no longer considered a distinct entity by the 2016 WHO CNS tumor classification, given its notable divergence between morphological features and molecular profiles; and its close resemblance to other diffuse gliomas (e.g., IDH mutant astrocytoma, IDH mutant and 1p/19q codeleted oligodendroglioma, and IDH wild-type glioblastoma). Widespread brain invasion involving three or more cerebral lobes, frequent bilateral growth, and regular extension to infratentorial structures is now mentioned as a special pattern of spread within the discussion of diffuse glioma subtypes [3]. This arises from the fact that the molecular characterization does not show any distinct subgroup of GC, but all GC can be assigned to the well-known molecular subgroup of gliomas [4]. It usually has bilateral involvement of the cerebral hemispheres and/or gray matter, with frequent extension to the brain stem, cerebellum, and less frequently to the spinal cord [5]. Strict adherence to radiological criteria is not recommended.

7.2 Epidemiology

The diffuse nature of GC with no specific symptom pattern makes it difficult to diagnose. Moreover, many cases are misdiagnosed and final confirmation of the diagnosis only comes on autopsy [6,7]. The available major case series compilations reveal a higher incidence in men compared to women, as is evident in the ANOCEF database of 296 cases [8]. A similar preponderance has been observed in the pediatric population [9]. No age group is exempt from the tumor as it has been identified from pediatric age groups as young as 4 months to the elderly well into the eighth decade [8,9].

7.3 Diagnosis

The diagnosis of GC poses a challenge at times due to the subtle and variable symptoms that represent the diffuse neuronal disruptions. Seizures are the most common presentation [6–8]. Other forms of presentation include focal deficits depending upon the area of involvement, manifestations of raised intracranial pressure, behavioral alterations, decreased academic performance, and decline in neurocognitive development mimicking dementia.

Misdiagnosis is common in view of clinical and radiological pictures that resemble various other conditions like viral encephalitis, acute disseminated encephalomyelitis, and vasculitis. Primary progressive multiple sclerosis in children should always have an alternative diagnosis of GC. Primary leukodystrophies affecting the white matter and heritable myelin disorders are other conditions to be ruled out.

When diagnosis of GC is suspected, it is important to have a biopsy, histological review, and also molecular characterization.

7.3.1 Imaging

The initial imaging modality is typically CT of the brain. CT does not usually reveal much except in cases with edema and mass effect. MRI findings are typically of a diffuse signal abnormality involving three or more contiguous cerebral lobes. Lesions appear iso- to hypointense on T1 with hyperintense on T2 (Figure 7.1). FLAIR imaging clearly delineates the extent of gliomatosis lesions due to the suppression of cerebrospinal fluid and is superior to conventional T2-weighted FSE images. In particular, the detection and delineation of cortical spread and the infiltration of the corpus callosum are best seen on FLAIR. Diffuse infiltrating type with no tumor mass formation

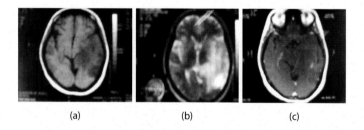

(a)   (b)   (c)

Figure 7.1   (a) T1-weighed magnetic resonance imaging of the brain showing an ill-defined hypointense lesion. (b) T2-weighted magnetic resonance imaging showing an area of diffuse, poorly defined, high signal intensity with a variable degree of obliteration of sulci and gyri. (c) Contrast magnetic resonance imaging showing mild diffuse patchy enhancement.

generally does not show any contrast enhancement. Methionine positron emission tomography (MET-PET) is used to differentiate GC from non-neoplastic lesions with similar findings on T2. Fluorodeoxyglucose positron emission tomography (FDG-PET) shows equivocal and nonspecific findings to have the definitive diagnosis.

7.3.2 Histopathology

The diagnosis of GC is a combination of imaging and histological findings. Biopsy can be performed by open or stereotactic method based on the mass effect and the presence of tumor mass in the image. In view of the diffuse nature of the tumor, single-location tumor sampling may not be the actual representation of the tumor pathology. Thus it is advisable to take two different biopsy samples using the same needle at different depths when using stereotaxic biopsy or at two different sites when using craniotomy as the procedure for tissue diagnosis. PET/MRI-guided biopsy can be helpful in identifying the relevant areas.

Histologically, the main characteristics include a diffuse pattern of neoplastic glial cells that grow with an absence of well-circumscribed margins. Large tumor cells with irregular elongated nuclei are evident (Figure 7.2). On a macroscopic level, the underlying architecture of brain tissue is attenuated, resulting in loss of gray and white matter distinction as cells infiltrate it diffusely but without a definable tumor mass. Infiltration of white matter usually respects the nervous fibers, although it can be associated with the destruction of myelin. Glial fibrillary acid protein (GFAP) and S-100 protein immunostaining varies from strong positivity to nonreactive. Proliferation correlates with grade, with Ki-67 range from 1% to 30% (Figure 7.3).

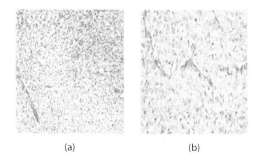

(a)   (b)

Figure 7.2   (a) Hematoxylin and eosin (H&E) 10× diffusely infiltrating small round to oval cells with pleomorphic nuclei. (b) H&E 40× gemistocytic astrocytes and scattered mitotic figures.

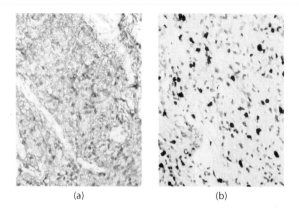

(a)   (b)

Figure 7.3   Immunohistochemistry of tumor cells showing (a) glial fibrillary acid protein positivity and (b) a Ki-67 labeling index of 40%.

GFAP expression suggests an astrocytic origin. The majority of cases are WHO Grade III astrocytomas, but oligodendrocytic origin and WHO Grades II and IV have been described [10].

7.3.3 Molecular alterations

GC has been linked to 1p/19q co-deletion and isocitrate dehydrogenase 1 (IDH1) somatic mutation; with IDH1R132H mutations being proposed as a favorable prognosis factor in adult studies. O6-methylguanine DNA methyltransferase promoter methylation is also found in GC, although its exact impact on prognosis has not been clearly determined.

The 2016 WHO CNS classification removes GC as a distinct entity, and considers it a growth pattern among various diffuse gliomas, including IDH-mutant astrocytic and oligodendroglial tumors as well as IDH–wild-type glioblastomas.

7.4 Treatment

The first surgery a patient may have is a biopsy to extract a sample for diagnosis. Most GC cases can be biopsied easily. As mentioned earlier, it is necessary to take a biopsy from two different areas at least a millimeter apart. In patients with mass effect and tumor mass formation, craniotomy and decompression of the tumor mass is the option. In cases of extensive infiltration in one lobe with a mass effect, a partial lobectomy with preservation of all eloquent areas can be considered to relieve the mass effect. The goal of surgery is to remove as much tumor as possible without damaging normal brain tissue in cases with mass effect and tumor formation.

Adjuvant radiotherapy and chemotherapy are indicated after surgery to slow disease progression [8]. Adjuvant radiotherapy protocol consists of radiotherapy to preoperative T2 FLAIR edema with margins to a total dose of 60 Gy, divided into 30 doses each of 2 Gy. It appears effective in delaying disease progression. Deterioration in quality of life and cognitive decline in children as a result of high doses and the wider areas to be covered are always a concern.

Concurrent chemotherapy is given with temozolomide 75 mg/m^2, followed by adjuvant temozolomide 150 mg/m^2 every 28 days for six cycles.

Novel therapeutics such as immunotherapies targeting programmed death ligand-1 and mutant IDH1 are in the developmental stage.

7.5 Prognosis

Median overall survival in adult series of 9.5–23.7 months and pediatric patient survival of 11–17 months have been reported [9]. Young age, good performance status, and low-grade histology have been identified as favorable prognostic factors. Alpha-internexin expression and IDH1R132H mutation are also considered favorable prognosis factors.

GC is a diffuse, aggressive type of glioma with a grim prognosis. Emerging molecular details leading to the development of novel therapeutics such as immunotherapies offer promise for treating GC in the future.

References

1. Nevin S. Gliomatosis cerebri. *Brain*. 1938;61:170–91.
2. Nevin S. Thalamic hypertrophy or gliomatosis of the optic thalamus. *J Neurol Psychiatry*. 1938;1(4):342–58.
3. Louis DN, Ohgaki H, Wiestler OD, et al. The 2007 WHO classification of tumours of the central nervous system. *Acta Neuropathol*. 2007;114:97–109.
4. Herrlinger U, Jones DTW, Glas M, et al. Gliomatosis cerebri: No evidence for a separate brain tumor entity. *Acta Neuropathol*. 2016;131(2):309–19.
5. Greenfield JP, Castañeda Heredia A, George E, Kieran MW, Morales La Madrid A. Gliomatosis cerebri: A consensus summary report from the First International Gliomatosis cerebri Group Meeting, March 26–27, 2015, Paris, France. *Pediatr Blood Cancer*. 2016;63(12):2072–7.

6. Rudà R, Bertero L, Sanson M. Gliomatosis cerebri: A review. *Curr Treat Options Neurol.* 2014;16:273–82.

7. Chen S, Tanaka S, Giannini C, et al. Gliomatosis cerebri: Clinical characteristics, management and outcomes. *J Neurooncol.* 2013;112(2):267–75.

8. Taillibert S, Chodkiewicz C, Laigle-Donadey F, Napolitano M, CartalatCarel S, Sanson M. Gliomatosis cerebri: A review of 296 cases from the ANOCEF database and the literature. *J Neurooncol.* 2006;76:201–5.

9. George E, Settler A, Connors S, Greenfield JP. Pediatric gliomatosis cerebri: A review of 15 years. *J Child Neurol.* 2016;31(3):378–87.

10. Sanson M, Napolitano M, Cartalat-Carel S, Taillibert S. La gliomatose cérébrale. *Rev Neurol (Paris).* 2005;161(2):173–81.

8
Oligodendroglioma

8.1 Definition

Oligodendroglioma (or oligodendroglial tumor) arises from oligodendrocytes in the central nervous system (CNS). Accounting for 10%–30%
of all gliomas, oligodendroglioma encompasses WHO Grade II oligodendroglioma (70%), which shows round and uniform nuclei with crisp
nuclear membranes, delicate chromatin, and small to inconspicuous
nucleoli; and WHO Grade III anaplastic oligodendroglioma (30%), which
displays enlarged and epithelioid cells with nuclei of increased size and
pleomorphism, a more vesicular chromatin pattern and more prominent
nucleoli [1].

The 2016 WHO classification of CNS tumors combines histological
criteria with molecular features and separates oligodendroglioma into
oligodendroglioma IDH mutant and 1p/19q codeleted (WHO Grade II), oligodendroglioma not otherwise specified (NOS) (WHO Grade II), anaplastic
oligodendroglioma IDH mutant and 1p/19q codeleted (WHO Grade III), and
anaplastic oligodendroglioma NOS (WHO Grade III) [1].

8.2 Biology

Glial cells (also called *neuroglia* or simply *glia*) are non-neuronal cells present
in both the CNS and the peripheral nervous system (PNS). Within the CNS
(i.e., the brain and spinal cord) four types of glial cells are found: astrocytes,
ependymal cells, microglia, and oligodendrocytes. On the other hand, inside
the PNS, two types of glial cells are noted: Schwann cells and satellite cells.

Oligodendrocytes are involved in the production of the myelin sheath that
insulates axons. Specifically, mature oligodendrocytes make myelin by wrap-
ping axons with their own cell membrane in a spiral shape, leading to the
formation of a multilayered sheath covering a long segment of axon. Myelin
membranes are rich in lipids, and cholesterol appears to be the rate-limiting
factor for myelin biogenesis. Although oligodendrocytes are morphologi-
cally distinct from the Schwann cells in the PNS, their axon–myelin units
are similar.

Affecting oligodendrocytes, oligodendroglioma mainly occurs in the supratentorial brain, involving the frontal lobe (50%), temporal lobe (35%), parietal lobe (7%), occipital lobe (1%–4%), cerebellum (3%), brainstem and spinal cord (1%), leptomeninges, cerebellopontine angle, cerebral ventricles, retina, and optic nerve. In patients experiencing generalized tonic–clonic seizures, oligodendroglioma often occurs in the mesial frontal regions (including the cortex connected to the genu of the corpus callosum). In contrast, in patients showing partial seizures, oligodendroglioma is located caudo-laterally in the orbitofrontal and temporal lobes, not in the cortex connected to the genu.

8.3 Epidemiology

As the second most common type of glioma after astrocytoma, oligodendroglioma represents 10%–30% of gliomas diagnosed. Oligodendroglioma may affect people of any age, with peak incidence in the fourth through sixth decades; pediatric cases are rarely encountered. There is a male predominance (male-to-female ratio of 2:1) among oligodendroglioma patients.

Grade III anaplastic oligodendroglioma is typically diagnosed in adults of 45–50 years, which is approximately 7–8 years older than those with Grade II oligodendroglioma. This age gap appears to reflect the average time of tumor evolution from Grade II to III.

8.4 Pathogenesis

Oligodendroglioma is linked to isocitrate dehydrogenase (*IDH*) gene mutation, codeletion of the short arm of chromosome 1 (1p) and the long arm of chromosome 19 (19q), O6-methylguanine-DNA methyltransferase (MGMT) promoter methylation, and alterations in glioma CpG island methylated phenotype (G-CIMP), homolog of *Drosophila capicua* (CIC), far-upstream binding protein 1 (FUBP1), telomerase reverse transcriptase (TERT), alpha thalassemia mental retardation syndrome X linked (ATRX), epidermal growth factor receptor (EGFR), and chromosome 7. Anaplastic oligodendroglioma often contains additional genetic aberrations, such as loss of heterozygosity at 9p and/or deletion of the *CDKN2A* gene (p16), *PIK3CA* mutations, and polysomies. Interestingly, EGFR amplification is mutually exclusive with 1p/19q codeletion and with IDH mutations, and its presence in histologically pure anaplastic oligodendrogliomas is indicative of small cell variant of glioblastoma (in 70% of cases). Compared to oligodendroglioma in adults, that in pediatric (childhood) and some young adulthood shows neither *IDH* mutation nor 1p/19q codeletion. Indeed, disseminated oligodendroglioma-like leptomeningeal neoplasms (which predominantly affect children) often harbor concurrent BRAF-KIAA1549 gene fusions and 1p deletion, without 19q deletion [2–5].

8.5 Clinical features

Clinical manifestations of oligodendroglioma include seizures (due to lesions in the frontal, parietal, and temporal lobes), headache, visual disturbances, and papilledema (due to increased intracranial pressure), weakness, cranial nerve palsy, irritability, apnea, hemiparesis, sensory neglect, and cognitive disorders.

8.6 Diagnosis

The initial diagnosis of oligodendrogliomas relies largely on CT and MRI assessment. While CT is more sensitive to calcification, MRI is superior for determining tumor extent and cortical involvement.

Oligodendroglioma is a relatively well-circumscribed (more circumscribed than astrocytoma), round to oval mass, with a characteristic affinity for the cortex (involving both the cortex and the subcortical white matter).

On imaging, oligodendroglioma is characterized by the presence of calcification and a cortical–subcortical location (most commonly in the frontal lobe, with focal thinning or remodeling of the overlying skull). Further, oligodendroglioma may show minimal to moderate enhancement and moderately increased perfusion. In pediatric oligodendroglioma, calcification, peritumoral edema and contrast enhancement are less common than in adults. Anaplastic oligodendroglioma may show new enhancement in a previously nonenhancing, untreated tumor in addition to increased perfusion, with edema, hemorrhage, cystic degeneration, and contrast enhancement more commonly seen in anaplastic oligodendroglioma than in pure oligodendroglioma [6].

On MRI, the tumor is typically hypointense on T1-weighted imaging, hyperintense on T2-weighted imaging, with a ringlike enhancement on postcontrast imaging and vascularization on perfusion imaging that correlates with grading and aggressiveness. Tumors with the 1p/19q codeletion more commonly show heterogeneous signal intensity on T2-weighted imaging, calcification, an indistinct margin, and mildly increased perfusion and metabolism than those with intact 1p/19q. On MR spectroscopy, the absence of a lactate/lipid peak distinguishes oligodendroglioma from anaplastic oligodendroglioma [6].

Differential imaging diagnosis of oligodendroglioma and anaplastic oligodendroglioma includes low-grade diffuse astrocytoma, ganglioglioma, dysembryoplastic neuroepithelial tumor, pleomorphic xanthoastrocytoma, and central neurocytoma. Whereas ganglioglioma, dysembryoplastic neuroepithelial tumor, and pleomorphic xanthoastrocytoma all have a similar cortical localization and ganglioglioma also shows calcification, they typically occur in a younger patient population than oligodendroglioma [6].

Histopathologically, oligodendroglioma (Grade II) contains uniform cells with round central nuclei, fine chromatin surrounded by a clear halo (or perinuclear halo, which is unstained cytoplasm resulting from an artifact of processing), delicate branching capillaries, and an absence of high-grade features. The combination of the round nuclei and perinuclear haloes results in a "fried egg" appearance of individual cells and in a honeycomb architectural pattern of groups of evenly spaced cells. Anaplastic oligodendroglioma (Grade III) typically shows focal or diffuse histological features of malignancy, with different degrees of malignant degeneration in the same lesion. Oligodendroglioma tends to calcify (which helps in radiological and histological diagnosis) and may form microcysts [6].

Histological differentiation between oligodendroglioma (WHO Grade II) and anaplastic oligodendroglioma (WHO Grade III) is essentially based on the criteria of cellularity, anaplasia, mitotic activity, microvascular proliferation, and necrosis. Oligodendroglioma displays a relatively low number of mitoses (5–10 per high-power field), moderate nuclear polymorphism, and minimal vascular endothelial proliferation. In contrast, anaplastic oligodendroglioma shows a higher degree of mitosis, nuclear polymorphism, vascular endothelial proliferation, and necrosis. When >80% of cells are oligodendroglial, the tumor is called *pure oligodendroglioma*. However, when >20% of cells are astrocytic, the tumor is called *oligoastrocytoma*.

Incorporation of both histological and molecular features (*IDH* and 1p/19q codeletion, *ATRX*) provides a more accurate diagnosis of oligodendroglioma. Oligodendroglioma is characterized by *IDH1/2* mutation, *ATRX* wild type, and 1p/19q loss. The testing processes often involve initial immunohistochemistry for ATRX and IDH1 R132H mutant protein, followed by 1p/19q analysis and subsequently by *IDH* sequencing (in case of positive ATRX staining of tumor cell nuclei and negative staining for IDH1 R132H, as *ATRX* mutations in *IDH* mutant diffuse gliomas almost never co-occur with 1p/19q codeletion and result in loss of immunohistochemical ATRX staining of tumor cell nuclei). Further assessment of *TERT* and/or *TP53* mutations may be exploited for establishing such a prognostic classification, canonical oligodendroglioma (which is typically *TERT* mutated and *TP53* wild type) [7].

In addition, assessment of the intrachromosomal ratio between loss of 1p centromere marker D1S2696, within the *NOTCH2* intron 12, and its 1q21 pericentric duplication N2N (N2/N2N test) provides a specific (100%) and sensitive (97%) approach to distinguish oligodendroglioma from glioblastoma.

8.7 Treatment

Maximal safe surgical resection followed by radiotherapy and chemotherapy is considered the standard treatment for oligodendroglioma. For anaplastic oligodendroglioma, postoperative radiotherapy is usually administered via external beam in standard fractions of 1.8–2 Gy, with a total dose in the range of 54–60 Gy. In contrast, chemotherapy often involves PCV (procarbazine, lomustine, and vincristine, also known as *procarbazine-CCNU-vincristine* or *PCV schedule*) or TMZ (temozolomide). TMZ is an oral alkylating agent that has ease of administration, good tolerability, and minimal myelotoxicity [8].

8.8 Prognosis

The median overall survival for patients with anaplastic oligodendroglioma is approximately 4.5 years. Patients treated with PCV followed by radiation have a median overall survival of 14.7 years, and those treated with only radiation have a median overall survival of 7.3 years. Patients with Grade II oligodendroglioma may survive for 11 years, but progression from Grade II oligodendroglioma to a high-grade malignant glioma may occur sooner or later.

Favorable clinical prognostic factors for oligodendrogliomas include younger age, good performance status, frontal lobe location, and presence of 1p/19q co-deletion. Patients with the combined allelic loss of chromosomes 1p and 19q often exhibit increased chemosensitivity and thus a better response to treatment. Indeed, the loss of 1p centromere marker D1S2696 within the *NOTCH2* intron 12 is found to have a good prognosis in oligodendroglioma. However, partial deletion of 1p or 19q lacks prognostic value.

Other genetic mutations may also influence the prognosis of oligodendrogliomas. For example, *IDH* mutant tumors without 1p/19q codeletion have a worse outcome compared to *IDH* mutant, 1p/19q codeleted Grade III tumors, but have a better than Grades II and III tumors without *IDH* mutations. *MGMT* methylation status is beneficial for alkylating chemotherapy in the absence of *IDH* mutations in both Grades III and IV gliomas. Oligodendrogliomas with *TERT* mutations in the presence of 1p/19q codeletion have a more favorable outcome than those with *TERT* mutations in the absence of 1p/19q codeletion or *IDH* mutations.

References

1. Louis DN, Ohgaki H, Wiestler OD. *WHO classification of tumours of the central nervous system*. 4th rev. ed. Lyon, France: IARC Press, 2016.
2. Alentorn A, Sanson M, Idbaih A. Oligodendrogliomas: New insights from the genetics and perspectives. *Curr Opin Oncol*. 2012;24(6):687–93.
3. Macaulay RJ. Impending impact of molecular pathology on classifying adult diffuse gliomas. *Cancer Control*. 2015;22(2):200–5.
4. Hofer S, Rushing E, Preusser M, Marosi C. Molecular biology of high-grade gliomas: What should the clinician know? *Chin J Cancer*. 2014;33(1):4–7.
5. Cahill DP, Louis DN, Cairncross JG. Molecular background of oligodendroglioma: 1p/19q, IDH, TERT, CIC and FUBP1. *CNS Oncol*. 2015;4(5):287–94.
6. Smits M. Imaging of oligodendroglioma. *Br J Radiol*. 2016;89(1060): 20150857.
7. Wesseling P, van den Bent M, Perry A. Oligodendroglioma: Pathology, molecular mechanisms and markers. *Acta Neuropathol*. 2015;129(6):809–27.
8. Simonetti G, Gaviani P, Botturi A, Innocenti A, Lamperti E, Silvani A. Clinical management of grade III oligodendroglioma. *Cancer Manag Res*. 2015;7:213–23.

9
Oligoastrocytoma

9.1 Definition

Oligoastrocytoma is a mixed glioma of the central nervous system (CNS) that contains both abnormal astrocytes and oligodendrocytes.

The 2016 WHO classification of CNS tumors separates oligoastrocytoma into two types: oligoastrocytoma, not otherwise specified (NOS; WHO Grade II) and anaplastic oligoastrocytoma, NOS (WHO Grade III), for which the molecular assessment is inconclusive or incomplete [1].

Oligoastrocytoma demonstrates histological features such as low to moderate cellularity, mild to moderate cytologic atypia, and occasional mitotic activity. In contrast, anaplastic oligoastrocytoma displays histological features of anaplasia/malignancy, including high cellularity, cellular pleomorphism, nuclear atypia, and high mitotic activity. Microvascular proliferation and necrosis may also be observed in anaplastic oligoastrocytoma but are not required for the diagnosis. Therefore, anaplastic oligoastrocytoma can be regarded as an oligoastrocytoma that grows quickly and has the potential to spread into surrounding brain tissue or to more distant parts of the body [1].

Because oligoastrocytoma NOS and anaplastic oligoastrocytoma NOS are unrecognized or undefined by current molecular diagnostic tests, in reporting, particular care should be exercised to eliminate possible technical problems (e.g., false-negative ATRX immunostaining or false-positive FISH results for 1p/19q codeletion due to regional differences within tissue specimens) that may contribute to misidentification [1].

Oligoastrocytoma that is detectable by molecular diagnostic tests is now classified as either astrocytoma (IDH mutant, ATRX mutant, and 1p/19q intact as well as TP53 mutant) or oligodendroglioma (IDH mutant, ATRX wild type, and 1p/19q codeleted, with 1p/19q-codeletion indicating whole-arm loss of the long arm of chromosome 1 and the short arm of chromosome 19) (see Chapters 2 and 8) [1,2].

9.2 Biology

Four types of glial cells (i.e., astrocytes, ependymocytes, microglia, and oligo-dendrocytes) exist in the CNS. Astrocytes are star-shaped glial cells involved in the provision of nutrients to neurons, maintenance of extracellular ion balance, and repair of traumatic injuries. Ependymocytes (or ependymal cells) are ciliated, columnar glial cells that line the cerebrospinal fluid–filled ventricles in the brain and the central canal of the spinal cord. Microglia are resident macrophages that utilize phagocytic and cytotoxic mechanisms to eliminate plaques, damaged neurons/synapses, and infectious agents in the CNS. Oligodendrocytes (or oligodendroglia) are branched, fried egg–shaped glial cells that form a covering layer for axons.

Whereas a tumor arising from astrocytes is known as an *astrocytoma* (constituting about 70% of reported brain tumors), a tumor that arises from ependymocytes is known as an *ependymoma* (<10%), and a tumor that arises from oligodendrocytes is known as an *oligodendroglioma* (about 20%). In addition, mixed glioma that involves both astrocytes and oligodendrocytes is referred to as *oligoastrocytoma* (about 2%) [3].

Oligoastrocytoma possibly originates *de novo* from one mother cell whose "offspring" may follow two slightly different developmental pathways (forming astrocytic and oligodendroglial cell populations) or through progression from a lower-grade tumor. As a supratentorial tumor, oligoastrocytoma is mainly found in the frontal (57%) and temporal (30%) lobes.

9.3 Epidemiology

Accounting for 5%–10% of all gliomas and 1% of all brain tumors, oligoastrocytoma mainly affects young and middle-aged adults, with peak incidence among people of 35–50 years (median age of 42.5 years at diagnosis). Very few children aged 0–14 years are detected with oligoastrocytoma [4,5].

9.4 Pathogenesis

Oligoastrocytoma that is classified by molecular diagnostic tests as astrocytoma is IDH mutant, alpha thalassemia mental retardation syndrome X linked (ATRX) mutant, and 1p/19q intact as well as TP53 mutant; that classified by molecular diagnostic tests as oligodendroglioma is IDH mutant, ATRX wild type, and 1p/19q codeleted. The presence of genetic mutations (e.g., deletions of chromosomes 1p and 19q and other changes) may lead to the production of unusual amounts of growth factors and gene proteins. This excess not only stimulates the aggressive growth of astrocytic and oligodendroglial cell populations to form tumor masses (which in turn may cause swelling or edema around the tumor and disrupt the normal function of these tissues through pressure and concussion) but also alters the functions of cells located elsewhere [6].

Oligoastrocytoma NOS and anaplastic oligoastrocytoma NOS may occur sporadically or progress from lower grade to higher grade. Nonetheless, the underlying tumorigenesis of these oligoastrocytomas and their progression remain unclear to date.

9.5 Clinical features

The most common symptoms associated with oligoastrocytoma are seizures, headaches, speech, motor, or personality changes. Other less common signs include weakness, numbness, or visual abnormality.

Oligoastrocytoma located in the frontal lobe of the brain may cause changes in the movement of arms and legs, personality and behavior characteristics, language, and ability to reason, leading to weakness on one side of the body, difficulty walking, seizures, difficulty remembering very recent occurrences, comments that do not match the conversation, or sudden changes in a person's usual behavior. In contrast, oligoastrocytoma located in the temporal lobe may induce changes in memory and the ability to understand language, interpret sensations, and comprehend visual images, leading to partial seizures and subtle language problems.

9.6 Diagnosis

After a neurological examination, patients suspected of oligoastrocytoma are subjected to CT and MRI of the whole brain and spinal cord. Further histological assessment of tumor tissue obtained through surgery or biopsy is necessary to confirm the diagnosis and grade of oligoastrocytoma.

Macroscopically, oligoastrocytoma resembles oligodendroglioma, which has no clear-cut borders, with the tumor cells intermingling with normal brain tissue. The tumor is soft and gray-pink in color, showing focally a mucoid, gelatinous matrix with occasional hemorrhage and/or calcification. The anaplastic form may display focally multiple smaller necroses. While Grade II oligoastrocytoma is slow growing, Grade III anaplastic oligoastrocytoma grows more quickly and is more aggressive. However, the rate of oligoastrocytoma growth may also depend on the proportion of astrocytes and oligodendrocytes in the tumor, since astrocytes appear to be more active than oligodendrocytes [3].

Classification of oligoastrocytoma into low-grade (WHO Grade II) or high-grade/anaplastic (WHO Grade III) forms relies on observation of histological features relating to cellularity, anaplasia, mitotic activity, microvascular proliferation, and necrosis.

Histologically, oligoastrocytoma shows the combined cytologic profiles of astrocytoma and oligodendroglioma. The former is characterized by large, irregular nuclei and many fibrillary cytoplasmic processes, whereas the latter is characterized by uniform, round nuclei with clear perinuclear halos (an artifact of fixation) and a fine, delicate capillary network (resembling chicken wire). In comparison with Grade II oligoastrocytoma, Grade III anaplastic oligoastrocytoma demonstrates increased cellularity, mitotic activity, nuclear atypia, and microvascular proliferation, with that displaying necrosis considered glioblastoma with an oligodendroglial component.

Further, oligoastrocytoma may be distinguished into astrocytoma or oligodendroglioma based on the presence or absence of chromosomes 1p and 19q. Indeed, up to 50% of oligoastrocytoma are found to have 1p and 19q codeletion and are subsequently classified and treated as oligodendroglioma. The status of loci 1p and 19q may be determined by FISH, loss of heterozygosity analysis, or virtual karyotyping. In particular, virtual karyotyping allows assessment of the entire genome (including the status of the 1p/19q loci, and the copy number of other key loci such as epidermal growth factor receptor or EGFR and TP53). The rare diffuse gliomas reported as oligoastrocytoma NOS and anaplastic oligoastrocytoma NOS are those in which histology does not allow for unequivocal categorization into either the astrocytic or oligodendroglial lineage tumors and in which molecular findings are inconclusive or incomplete [1].

9.7 Treatment

For Grade II oligoastrocytoma, surgical biopsy and removal of as much the tumor as possible without causing severe neurological damage or disturbing eloquent regions of the brain (speech/motor cortex) represent the initial treatment options. Focal fractionated radiation therapy may help control seizures related to the tumor that are not controllable with seizure medication. For recurrent tumor, surgical resection, radiation (if not previously radiated), and chemotherapy may be considered [7].

For Grade III oligoastrocytoma, surgical biopsy and removal are also used as initial treatments. Focal fractionated radiation and chemotherapy often start 2–4 weeks after surgery [7].

Oligoastrocytoma containing deletion of chromosomes 1p or 19q may be treated with oral chemotherapy (e.g., temozolomide [Temodar] or PCV [procarbazine, lomustine (CCNU), and vincristine]). Targeted therapy consists of bevacizumab (Avastin). Steroids may be used to decrease edema associated with brain tumors (which often cause swelling or edema in surrounding tissues). Antiepileptic drugs or anticonvulsant drugs may be used to control seizures. Antiemetic drugs may be used to prevent vomiting and control nausea [7,8].

9.8 Prognosis

Prognosis of oligoastrocytoma depends on the age of the patient, location of the tumor, grade of the tumor cells, status of chromosomes 1p and 19q, and amount of tumor removed during surgery, as well as adoption of postoperative radiotherapy and/or chemotherapy [9,10].

In general, younger age (<40 years), low-grade initial diagnosis, and better extent of resection are factors in improved survival time. Oligoastrocytoma harboring 1p/19q codeletion (i.e., oligodendroglioma) responds well to the PCV chemotherapy and radiation therapy. Median survival for Grades II and III oligodendroglioma is about 11 and 4–5 years, respectively, but shorter for oligoastrocytoma.

References

1. Louis DN, Ohgaki H, Wiestler OD. *WHO classification of tumours of the central nervous system*. 4th rev. ed. Lyon, France: IARC Press, 2016.
2. Wesseling P, van den Bent M, Perry A. Oligodendroglioma: Pathology, molecular mechanisms and markers. *Acta Neuropathol*. 2015; 129(6): 809–27.
3. Walker C, Baborie A, Crooks D, Wilkins S, Jenkinson MD. Biology, genetics and imaging of glial cell tumours. *Br J Radiol*. 2011; 84(2): S90–106.
4. Forst DA, Nahed BV, Loeffler JS, Batchelor TT. Low-grade gliomas. *Oncologist*. 2014; 19(4): 403–13.
5. Hofer S, Rushing E, Preusser M, Marosi C. Molecular biology of high-grade gliomas: What should the clinician know? *Chin J Cancer*. 2014; 33(1): 4–7.

6. Cahill DP, Louis DN, Cairncross JG. Molecular background of oligo-dendroglioma: 1p/19q, IDH, TERT, CIC and FUBP1. *CNS Oncol.* 2015; 4(5): 287–94.

7. Khan KA, Abbasi AN, Ali N. Treatment updates regarding anaplastic oligodendroglioma and anaplastic oligoastrocytoma. *J Coll Physicians Surg Pak.* 2014; 24(12): 935–39.

8. Lecavalier-Barsoum M, Quon H, Abdulkarim B. Adjuvant treatment of anaplastic oligodendrogliomas and oligoastrocytomas. *Cochrane Database Syst Rev.* 2014; (5): CD007104.

9. Riemenschneider MJ, Reifenberger G. Molecular neuropathology of gliomas. *Int J Mol Sci.* 2009; 10(1): 184–212.

10. Suvà ML. Genetics and epigenetics of gliomas. *Swiss Med Wkly.* 2014; 144: w14018.

10
Ependymoma

10.1 Definition

Arising from the wall of the ventricular system along the craniospinal axis (or neuroaxis), ependymal tumor (or ependymoma) is separated by the 2016 WHO classification of central nervous system (CNS) tumors into five clinically and genetically distinct subtypes: (i) subependymoma (WHO Grade I), (ii) myxopapillary ependymoma (WHO Grade I), (iii) ependymoma (papillary, clear cell, and tanycytic) (WHO Grade II), (iv) ependymoma *RELA* fusion–positive (WHO Grade II/III), and (v) anaplastic ependymoma (WHO Grade III) [1–3].

Based on DNA methylation, gene expression profiles, and the spectrum of genomic alterations, ependymoma may be differentiated into three biological subtypes: (i) infratentorial tumor (posterior fossa A, CpG island methylator phenotype [CIMP]-positive ependymoma, termed *EPN-PFA*, and posterior fossa B, CIMP-negative ependymoma, termed *EPN-PFB*]; (ii) supratentorial tumor (*C11orf95-RELA*–positive ependymoma, and *C11orf95-RELA*–negative and *YAP1* fusion–positive ependymoma); and (iii) spinal tumor. This classification with a distinct molecular association offers new insight into the prognosis for different types of ependymoma [2].

10.2 Biology

Ependymal cells (or ependymocytes) are one of the four glial cell types (i.e., astrocytes, ependymal cells, microglia, and oligodendrocytes) in the CNS. Lining the ventricles (cerebrospinal fluid or CSF-filled spaces) of the brain and the central canal (CSF-filled space down the center) of the spinal cord, ependymal cells form a ciliated, columnar structure (the ependyma), with resemblance to mucosal epithelium.

Ependymoma evolves from the ependymal cells (with regular, round to oval nuclei and gland-like round or elongated structures that extend into the lumen) in different parts of the neuroaxis, typically the posterior fossa (the area of the brain below the tentorium, containing the cerebellum and brainstem), the supratentorium (the area of the brain above the tentorium containing the cerebral hemispheres), and the spinal cord.

Because ependymoma-initiating cells harbor specific markers of radial glial progenitor cells such as BLBP or RC2 and show Notch pathway activation, it is possible that ependymoma may also derive from radial glia cells.

Over 75% of ependymoma in adults are found within the spinal canal. In children, about 90% of ependymoma are detected within the brain in the posterior fossa, in or around the fourth ventricle (situated in the lower back portion of the brain), and only 10% occur within the spinal cord. In addition, ependymoma may form in the choroid plexus (tissue in the ventricles that makes cerebrospinal fluid) or rarely in the pelvic cavity [2].

The most common posterior fossa ependymoma subtype is EPN-PFA, which is characterized by its presence in young children (median age of 3 years); its low rates of mutation that affect protein structure (approximately five per genome), with no recurring mutations; and its balanced chromosomal profile with few chromosomal gains or losses.

10.3 Epidemiology

Affecting both children (of 3–6 years) and adults (peak onset at 30 years), ependymoma represents the third most common form of childhood brain and spine tumors and accounts for about 10% of childhood CNS tumors and about 5% of adult intracranial gliomas. There is a notable male predominance among ependymoma patients. Ependymoma has a tendency to recur at the primary tumor site and is associated with neurofibromatosis Type 2 and syringomyelia [1].

10.4 Pathogenesis

Usually found in the floor of the fourth ventricle (situated in the lower back portion of the brain) as well as the spinal cord, conus medullaris, and supratentorial locations, ependymoma may obstruct the flow of cerebrospinal fluid, induce hydrocephalus, and alter the developmental expression profiles of anatomically restricted progenitor cells in the CNS.

Molecularly, ependymoma is linked to loss of heterozygosity (LOH) in chromosomes 1p, 6q, 9, 11q (with a significant inverse correlation with LOH of 22q), 16, 17, 19q, and 22q and gain in chromosomes 1q, 5p, and 7. Interestingly, the allelic loss on 22q tends to occur in intracranial anaplastic ependymoma in children and in intraspinal ependymoma in adults, whereas chromosomal gain of 1q is frequently seen in pediatric patients, patients with intracranial lesions, and in Grade III ependymoma [2,4,5].

Specific prognostic markers or potential therapeutic targets in myxopapillary ependymoma are homeobox (HOX) protein HOXB13 and platelet-derived growth factor receptor alpha. Those in subependymoma are the ETV6, YWHAE, TOP2A, TLR2, IRAK1, TIA1, and UFD1L genes; those in clear cell ependymoma are loss of chromosome 9 and aberrations in chromosomes 1q and 3. Those in anaplastic ependymoma are gain of 1q (usually in the posterior fossa) and loss of 9 [6].

Biomarkers with implication for poor prognosis include 1q gain ±9p21.3 loss for both adult and pediatric ependymomas, *EGFR* overexpression, and *EVI-1* expression for adult ependymomas, as well as expression of the *C11orf95–RELA* gene fusion within chromosome 11q (which is possibly caused by chromothripsis) in supratentorial ependymoma [2].

10.5 Clinical features

Ependymoma may induce headache, nausea/vomiting, loss of appetite, pain (or stiffness) in the neck and back, weakness in the legs, gait change (rotation of feet when walking), blurry vision (seeing vertical or horizontal lines when in bright light, temporary inability to distinguish colors, and visual loss), change in bowel function (impaction/constipation), trouble urinating, confusion or irritability, sleeping difficulty or drowsiness, uncontrollable twitching, back flexibility, seizures, and temporary memory loss.

Specifically, children with posterior fossa ependymoma often have headache, vomiting, ataxia, neck pain, double vision, or cranial nerve palsies. Patients with supratentorial ependymoma may display headache, seizures, or location-dependent focal neurologic deficits. Patients with spinal cord ependymoma (usually the myxopapillary variant) may show lower back pain, sciatica, extremity weakness, leg length discrepancy and scoliosi, and bowel and bladder dysfunction.

10.6 Diagnosis

Subependymoma (WHO Grade I) is a rare, slow-growing, and well-circumscribed tumor in or around the ventricles, without extending into the brain parenchyma. The tumor mainly affects males of >40 years.

Myxopapillary ependymoma (WHO Grade I) is a benign, slow-growing, well-circumscribed tumor located in the base (the filum terminale) of the spine. Rarely seen in children, it accounts for about 10% of ependymal tumors.

Ependymoma (WHO Grade II) is a common, fairly well delineated tumor in the walls of the ventricles or the spinal canal. The tumor shows peak occurrence at 5 years and 35 years and consists of three variants/subtypes: papillary (the surfaces with cerebrospinal fluid exposure), clear cell (the supratentorial compartment of the brain), and tanycytic (the spinal cord).

Ependymoma *RELA* fusion–positive (WHO Grade II/III) is phenotypically similar to ependymoma or anaplastic ependymoma, and characterized by a *C11orf95-RELA* fusion. It accounts for the majority of supratentorial ependymal tumors in children.

Anaplastic ependymoma (or malignant ependymoma) (WHO Grade III) is a fast-growing, irregularly lobulated or well-demarcated and quasi-circular mass of 3.0–7.3 cm in size, located in the lower back part of the skull (posterior fossa) of children and the brain of adults. It often displays cysts, necrosis, hemorrhage, and vascular proliferation. On T1-weighted imaging, anaplastic ependymoma is iso-hypointense relative to the gray matter, with additional small- or medium-sized foci of hyperintensity. On T2-weighted imaging, anaplastic ependymoma is heterogeneous (with areas of hypointense and hyperintense signal intensity) or slightly hypointense relative to the gray matter. Contrast-enhanced MRI shows areas of marked (wreath-like or ring-like) or mild (flake-like) enhancement. Certain patchy areas of hypointensity have no contrast enhancement. Cystic or necrotic areas are frequently observed, and peritumoral edema is moderate, mild, or absent.

Major histological criteria for diagnosing ependymoma include (i) observation of perivascular pseudorosettes, which are composed of nucleus-free mantles surrounded by a radial disposition of cells around blood vessels; (ii) presence of true ependymal rosettes and ependymal canals, consisting of columnar cells arranged around a central lumen or cavity; (iii) immunohistochemical detection of strong cytoplasmic positivity for glial fibrillary acidic protein (GFAP), vimentin, and CD56 [1].

Under the microscope, anaplastic ependymoma demonstrates plentiful small, round, or fusiform tumor cells with large and polymorphic nuclei and scant cytoplasm; perivascular pseudorosettes; and increased cellularity, brisk mitotic activity, microvascular proliferation, and pseudopalisading necrosis. A positive Ki-67 labeling index is more common in anaplastic ependymoma than in low-grade ependymoma. The tumor stains positive for GFAP and p53 protein, variable for vimentin and S-100 protein, and negative or weakly positive for epithelial membrane antigen (EMA) [1].

Application of molecular techniques enables improved definition of ependymoma. EPN-PFA (which occurs primarily in young children) is shown to have a largely balanced genomic profile with an increased occurrence of chromosome 1q gain and expression of genes and proteins (e.g., tenascin C and epidermal growth factor receptor). In contrast, EPN-PFB (which occurs primarily in older children and adults) is shown to have numerous cytogenetic abnormalities involving whole chromosomes or chromosomal arms [1].

Differential diagnosis for ependymoma includes pilocytic astrocytoma, oligodendroglioma, hemangioblastoma, and glioblastoma. Often occurring in children, pilocytic astrocytoma has a large cystic component, with small enhancing mural nodules but without peritumoral edema. Oligodendroglioma is rarely cystic and shows mild enhancement. Hemangioblastoma is a solid cystic mass, with multiple voids. Glioblastoma often grows contralaterally across the midline and involves the bilateral frontal lobes, and tends to display more heterogeneous intensity and more marked peritumoral edema [1].

10.7 Treatment

Surgical removal offers an effective treatment for well-differentiated ependymoma (e.g., subependymoma and myxopapillary ependymoma), although surgery along with adjuvant radiotherapy and chemotherapy is necessary for other ependymoma (e.g., posterior fossa ependymoma and anaplastic ependymoma). About 50%–70% of childhood ependymomas are cured with surgery and irradiation, but some may recur. For recurrent ependymoma, chemotherapy with carboplatin, cisplatin, cyclophosphamide, etoposide, lomustine, methotrexate, and vincristine may be prescribed [7,8].

10.8 Prognosis

Prognosis of ependymoma depends on a number of factors, including the location and type of tumor, possible changes in the genes or chromosomes, presence of residual cancer cells after tumor removal, spreading to other parts of the brain or spinal cord, and age of affected child.

In general, ependymoma in young children (<4 years) has a poor prognosis due to difficulty in surgery and inherent chemo- and radio-resistance, with about half of the patients succumbing to the disease. EPN-PFA in young children often produces a less favorable outcome than EPN-PFB in older children and adults. Cranial ependymoma has a worse prognosis than primary spinal cord ependymoma; ependymoma in the lower portion of the spinal cord has a worse prognosis than that in the lower portion.

Potent prognostic predictors for ependymoma include telomerase (telomerase reverse transcriptase), EGF receptor, intratumoral immune response, topoisomerase II, TP53, nestin, and tenascin-C, the overexpression of which will negatively impact patient recovery.

Up to 60% of children with intracranial ependymoma, mostly represented by either classic (WHO Grade II) or anaplastic (WHO Grade III) tumors, will survive for >5 years.

References

1. Jain A, Amin AG, Jain P, et al. Subependymoma: Clinical features and surgical outcomes. *Neurol Res.* 2012;34(7):677–84.
2. Benson R, Mallick S, Julka PK, Rath GK. Molecular predictive and prognostic factors in ependymoma. *Neurol India.* 2016;64(2):279–86.
3. Louis DN, Ohgaki H, Wiestler OD. *WHO classification of tumours of the central nervous system.* 4th rev. ed. Lyon, France: IARC Press, 2016.
4. Yao Y, Mack SC, Taylor MD. Molecular genetics of ependymoma. *Chin J Cancer.* 2011;30(10):669–81.
5. Nobusawa S, Hirato J, Yokoo H. Molecular genetics of ependymomas and pediatric diffuse gliomas: A short review. *Brain Tumor Pathol.* 2014;31(4):229–33.
6. Andreiuolo F, Ferreira C, Puget S, Grill J. Current and evolving knowledge of prognostic factors for pediatric ependymomas. *Future Oncol.* 2013;9(2):183–91.
7. Khatua S, Ramaswamy V, Bouffet E. Current therapy and the evolving molecular landscape of paediatric ependymoma. *Eur J Cancer.* 2016;70:34–41.
8. PDQ Pediatric Treatment Editorial Board. *Childhood Ependymoma Treatment (PDQ®): Health Professional Version.* PDQ Cancer Information Summaries. Bethesda, MD: National Cancer Institute (US), 2002.

11
Choroid Plexus Tumors

11.1 Definition

Choroid plexus tumors are rare intraventricular neoplasms that encompass three histological subgroups: choroid plexus papilloma (CPP), atypical choroid plexus papilloma (ACPP), and choroid plexus carcinoma (CPC) [1].

- *CPP (WHO Grade I)* is a benign tumor with single or multiple layers of cuboidal to columnar epithelial cells covering fibrovascular connective tissue cores. Tumor cells display bland cytologic features with uniform nuclei and low-grade polymorphism.
- *ACPP (WHO Grade II)* is a benign tumor characterized by increased mitotic activity (>2 mitoses per 10 high-power fields) and occasional focal blurring of papillary pattern, necrosis, and invasion of CNS. As a histologic and biological intermediate between CPP (Grade I) and CPC (Grade III), ACPP presents a diagnostic challenge.
- *CPC (WHO Grade III)* is a malignant tumor with frankly aggressive histologic features (e.g., increased cellularity, disarrangement of papillary architecture, high mitotic activity, nuclear pleomorphism, invasion of CNS tissue, and necrosis), variable clinical behavior, high rates of relapse, and poor outcome.

11.2 Biology

Choroid plexus tumors are primary brain tumors originating from the epithelial cells of the choroid plexus (which lines the ventricles of the brain and produces cerebrospinal fluid or CSF). Although CPP, ACPP, and CPC typically arise *de novo*, some ACPP and CPC may result from malignant transformation of existing CPP.

As noncancerous neoplasms, CPP and ACPP grow slowly and rarely spread to other parts of the brain or spinal cord. In contrast, CPC is a fast-growing, highly aggressive malignant tumor with the tendency to spread through the CSF (spinal drop metastases) and by leptomeningeal dissemination to nearby tissue.

Choroid plexus tumors occur more frequently in the supratentorial region than the infratentorial region. In children (median age of 1.5 years), choroid plexus tumors are found in the lateral ventricles (50%), fourth ventricle (40%), third ventricle (5%), and multiple ventricles (5%), whereas in adults aged 20–35 (median age of 22.5 years), these tumors are mainly present in the fourth ventricle and cerebellopontine angle.

11.3 Epidemiology

Choroid plexus tumors are rare intraventricular papillary neoplasms, accounting for approximately 2%–4% of intracranial tumors in children (usually <2 years of age, especially neonates) and 0.5% in adults.

While benign CPP and ACPP are responsible for 80%–90% of all choroid plexus tumors, malignant CPC represents 10%–20% of such tumors (mostly in children).

11.4 Pathogenesis

Although a majority of choroid plexus tumors arise sporadically, some are linked to hereditary factors. CPP appears to be a component of Aicardi syndrome and may also develop in the context of Down syndrome, von Hippel–Lindau disease, and neurofibromatosis type II. CPC is associated with Li–Fraumeni (LFS) and rhabdoid predisposition syndromes, with germline mutations in tumor-suppressor genes *TP53* and *hSNF5/INI1/SMARCB1*, respectively. Additionally, somatic mutations in *TP53* are identified in 60% of CPC.

Focal chromosomal gains related to choroid plexus tumors include chromosomes 14q21–q22 (*OTX2*), 7q31.1 (*LAMB1*), 9q21.12 (*TRPM3*), and 20p12 (*PLCB1*). Additional focal alterations are identified in papillomas (e.g., gains of chromosomes 5q, 6q, 7, 8, 9, 11, 12, 15, 17, 18, 19, 20, and 21 and losses of chromosomes 2, 10, 13q, and 21q), in atypical papillomas (gains of chromosomes 5, 7, 8, 9, 11, 12, and 20, but not chromosomal losses), and in carcinomas (e.g., gains of chromosomes 1, 4, 8q, 9p, 12, 14q, 20q, and 21 and losses of chromosomes 3p, 5, 9q, 10q, 13q, 18q, 22q) [2].

Specific gene alterations in CPP include *ARL4A* (7p21.3) and those encoding Notch receptors (Notch 1, 2, and 3). Specific gene mutations involved in CPC include *RAB6B, C3orf36, SLCO2A1,* and *RYK* (3q22); *POLH, GTPBP2, MAD2L1BP, MRPS18A, RSPH9,* and *VEGFA* (6p21); *RBFOX1* (16p13.3); and *PDGF*. Several oncogenes (e.g., *Taf12, Nfyc,* and *Rad54l*) have also been implicated in the initiation and maintenance of CPC, as overexpression of these genes accelerates CPC tumorigenesis in cooperation with deletion of *Trp53, Rb,* and *Pten* [2].

11.5 Clinical features

Clinical presentation of choroid plexus tumors is largely attributable to hydrocephalus and increased intracranial pressure caused by tumor mass–related obstruction of normal CSF flow, overproduction of CSF, local expansion of the ventricles, or spontaneous hemorrhage, with symptoms ranging from headache, nausea, vomiting, papilledema, ataxia, diplopia, strabismus, increased head circumference, bulging fontanelles (the soft spot at the top of the skull in infants), and developmental delay to altered mental status.

11.6 Diagnosis

Choroid Plexus Papilloma (CPP) is a lobular, solid, well-circumscribed, intraventricular, cauliflower-like mass adhering to the ventricular wall but without invading into the brain parenchyma proper. Calcification, cyst, and hemorrhage may be observed. On MRI, CPP appears homogeneous and isointense with respect to gray matter on T1 images and hyperintense on T2 images, and exhibits intense contrast enhancement. Microscopically, CPP contains fibrovascular papillary projections that are lined by cuboidal to columnar epithelium with cell crowding, elongation, and stratification compared to the orderly cobblestone appearance of normal choroid plexus tissue. CPP may show eosinophilic or clear cytoplasm and bland nuclei with fine chromatin. Occasionally CPP may display oncocytic change, melanization, cytoplasmic vacuolization, tubular–glandular architecture, focal ependymal differentiation, or neuropil-like islands, along with angioma-like increase in blood vessels, hyalinization or calcification, xanthomatous or mucinous change, or formation of metaplastic bone, cartilage, or adipose tissue. CPP consistently stains positive for pancytokeratin and vimentin; variably positive for S-100 protein (54%), glial fibrillary acidic protein (GFAP, 70%), transthyretin (89%), epithelial membrane antigen (EMA, 11-77%), and synaptophysin (occasionally); but negative for p53 (usually), CK20 (usually), BerEP4, and carcinoembryonic antigen (CEA) [3,4].

Atypical Choroid Plexus Papilloma (ACPP) often has an irregular margin with adjacent white matter edema and shows increased mitotic activity (defined as ≥2 mitoses per 10 high-power fields), cribriforming, anastomosing papillary formations, focal solid growth patterns, hypercellularity, nuclear pleomorphism, and necrosis [5].

Choroid Plexus Carcinoma (CPC) is a solid mass with invasion into the periventricular brain parenchyma, hemorrhage, calcification, and necrosis.

On non-contrast CT, CPC is heterogeneous and typically iso- to hyperdense to grey matter, with calcification in 20%–25% of cases, and prominent but heterogeneous contrast enhancement due to areas of necrosis and cyst formation. On MRI, CPC shows a heterogeneous appearance with areas of necrosis, calcification, or hemorrhage, and appears iso- to hypointense on T1, iso- to hypointense with hyperintense necrotic areas on T2, blooming from calcifications/hemorrhage on T2* GRE, and marked, heterogeneous enhancement on T1 C+ (Gd). Microscopically, CPC is noted for its features of frank malignancy such as elevated mitotic activity (>5 mitoses per 10 high-power fields), hypercellularity, nuclear pleomorphism, solid growth with sheets of tumor cells, and necrosis as well as extensive invasion into the surrounding brain parenchyma. CPC may also have tight three-dimensional clusters and isolated anaplastic cells, a high nuclear-to-cytoplasmic ratio, nuclear indentations and lobulations, single or multiple micronucleoli, and scant, pale, granular cytoplasm. CPC is generally positive for cytokeratin, EAAT1, Ki17.1, and stanniocalcin-1; focally reactive for synaptophysin, GFAP (20%), CA19-9, EMA, and CEA; and variably positive for S-100 and trans-thyretin. CPC also shows nuclear positivity for p53 and INI-1 and metastatic CPC is positive for HEA-125 and BerEP4 [6,7].

Differential diagnosis of CPP includes both normal choroid plexus and villous hypertrophy of the choroid plexus, papillary ependymoma (negative for laminin and collagen IV as well as E-cadherin), and astroblastoma. Differential diagnosis of CPC includes anaplastic ependymoma and atypical teratoid/rhabdoid tumor (AT/RT, which is negative for INI1) [7].

11.7 Treatment

Gross total resection is usually curative for CPP and ACPP; adjuvant radiotherapy and/or chemotherapy may be required for CPC and incompletely resected ACPP.

Tumor removal helps relieve hydrocephalus (excess water in the brain) in half of patients, whereas shunt (tube or drainage system) does so in other patients. Operative complications for choroid plexus tumors include pneumocephalus (40%), focal deficits (36%), subdural effusion (32%), and persistent hydrocephalus requiring shunt (24%).

Radiotherapy and/or chemotherapy (e.g., ifosfamide, cisplatin, etoposide, cyclophosphamide, vincristine, carboplatin, bevacizumab, temozolomide, and intrathecal methotrexate) is recommended for patients with incompletely removed or nonremovable tumor. A second surgery together with radiotherapy and/or chemotherapy is usually suggested for recurrent tumor [8–10].

11.8 Prognosis

With an indolent clinical behavior, CPP has a favorable long-term, postsurgery survival rate (>90%). The 5-year overall survival rates for CPP can reach 90%–100% after gross total resection or even partial resection.

ACPP has a 5-year overall survival rate of 89% after surgery but may carry a greater risk of local recurrence than CPP, especially in patients with incompletely resected tumor. Prognosis worsens in patients with ACPP showing decreased S-100 protein expression (<50% of cells strongly positive for S-100), lack of immunoreactivity for transthyretin, brain invasion, lack of marked stromal edema, and presence of necrosis.

CPC carries an extremely poor prognosis, with a 5-year overall survival rate of 36% after surgery and long-term cognitive and developmental deficits in surviving patients. CPC showing TP53 mutation (two copies worse than one copy), brain invasion, CSF seeding, and leptomeningeal dissemination has a less favorable outcome, whereas that harboring chromosome 9p gain or 10q loss has a more favorable clinical course. The outcome of CPC improves when gross total resection is combined with adjuvant chemotherapy and/or local radiotherapy.

References

1. Louis DN, Perry A, Reifenberger G, et al. The 2016 World Health Organization classification of tumors of the central nervous system: A summary. *Acta Neuropathol*. 2016;131:803–20.
2. Merino DM, Shlien A, Villani A, et al. Molecular characterization of choroid plexus tumors reveals novel clinically relevant subgroups. *Clin Cancer Res*. 2015;21(1):184–92.
3. Mishra A, Ojha BK, Chandra A, Singh SK, Chandra N, Srivastava C. Choroid plexus papilloma of posterior third ventricle: A case report and review of literature. *Asian J Neurosurg*. 2014;9(4):238.
4. Shenoy AS, Desai HM. Choroid plexus carcinoma with hyaline globules: An unusual histological feature. *Indian J Pathol Microbiol*. 2016;59(2):249–50.
5. Pandey S, Sharma V, Singh K, Ghosh A, Gupta PK. Uncommon presentation of choroid plexus papilloma in an infant. *J Pediatr Neurosci*. 2016;11(1):61–3.

6. Ozdogan S, Gergin YE, Gergin S, et al. Choroid plexus carcinoma in adults: An extremely rare case. *Pan Afr Med J.* 2015;20:302.

7. Dangouloff-Ros V, Grevent D, Pagès M, et al. Choroid plexus neoplasms: Toward a distinction between carcinoma and papilloma using arterial spin-labeling. *AJNR Am J Neuroradiol.* 2015;36(9):1786–90.

8. Bettegowda C, Adogwa O, Mehta V, et al. Treatment of choroid plexus tumors: A 20-year single institutional experience. *J Neurosurg Pediatr.* 2012;10(5):398–405.

9. Kamar FG, Kairouz VF, Nasser SM, Faddoul SG, Saikali IC. Atypical choroid plexus papilloma treated with single agent bevacizumab. *Rare Tumors.* 2014;6(1):4687.

10. Bohara M, Hirabaru M, Fujio S, et al. Choroid plexus tumors: Experience of 10 cases with special references to adult cases. *Neurol Med Chir (Tokyo).* 2015;55(12):891–900.

12
Dysembryoplastic Neuroepithelial Tumor

12.1 Definition

As a heterogeneous group of neoplasms affecting the neuroepithelial tissue, neuronal and mixed neuronal-glial tumors range from Grade I (desmoplastic infantile astrocytoma and ganglioglioma, dysembryoplastic neuroepithelial tumor, dysplastic gangliocytoma of the cerebellum—Lhemitte–Duclos disease, gangliocytoma, ganglioglioma, papillary glioneuronal tumor, paraganglioma of the filum terminale, rosette-forming glioneuronal tumor of the fourth ventricle), Grade II (central neurocytoma, extraventricular neurocytoma, cerebellar liponeurocytoma), and Grade III (anaplastic ganglioglioma) to grades unknown (diffuse leptomeningeal glioneuronal tumor) [1]. These tumors all contain cells of neuronal and sometimes glial differentiation and may be associated with dysplasia, hamartoma, and other malformations. The most common neuronal and mixed neuronal-glial tumors are ganglioglioma and dysembryoplastic neuroepithelial tumor, which account for about 70% and 20% of cases belonging to the group, respectively.

Dysembryoplastic neuroepithelial tumor (DNET) is a benign neuronal and mixed neuronal–glial neoplasm (WHO Grade I) noted for its supratentorial intracortical location, multinodular architecture, and heterogeneous cellular composition.

Clinically, DNET is associated with pharmaco-resistant chronic epilepsy, with partial complex seizures as the main presentation. Radiologically, DNET is characterized by a cortical topography and lack of mass effect or perilesional edema. Histologically, DNET demonstrates a specific glioneuronal (GN) element and association with glial nodules and focal cortical dysplasia (FCD), variations of which permit its differentiation into complex (or typical, solitary nodular), simple (multinodular), and nonspecific (diffuse) histologic forms/subtypes [1].

The complex form of DNET is characterized by the presence of specific GN elements together with glial nodules and/or FCD; the simple form of DNET consists of only specific GN elements. The nonspecific form of DNET lacks specific GN elements but has glial nodules that are similar to those observed in complex DNET [1].

12.2 Biology

DNET typically arises from its supratentorial intracortical location (especially the temporal lobe and occasionally the caudate nucleus, septum pellucidum, brainstem, and cerebellum).

Based on its mixed cellularity, preponderance in the temporal lobe, and association with FCD, DNET may have a developmental origin from the secondary germinal layer and pluripotent precursor cells. In addition, with higher expression levels of nestin, MAP2, and CD34 stem cell markers, the nonspecific DNET form appears to have an earlier developmental origin than simple or complex DNET forms.

Although DNET has been generally described as having a benign course with cortical dysplasia rather than true neoplasia, there are occasional reports of malignant transformation in DNET.

12.3 Epidemiology

DNET is a rare glioneuronal tumor that tends to present with medically intractable epileptic seizures in childhood, adolescence, or young adulthood (with a range of 6–20 years and a slight male predominance). The nonspecific form of DNET is mainly found in adults.

12.4 Pathogenesis

A number of factors have been implicated in the almost ubiquitous presence of seizures with brain tumors like DNET and ganglioglioma. These include differential alterations in regional metabolism and pH, immunologic activity, disordered neuronal function, altered vascular supply and permeability, release of altered tumoral molecules (amino acids, proteins, and enzymes), and abnormal protein transport and binding to receptors.

Whole-exome sequencing and targeted sequencing have uncovered the association of germline *FGFR1* mutation (p.R661P) and somatic activating *FGFR1* mutations (p.N546K or p.K656E) as well as MAP kinase pathway activation with DNET. *FGFR1* alterations mainly comprise intragenic tyrosine kinase FGFR1 duplication and multiple hotspot mutations *in cis*. Depending on the type and localization of the mutation along the gene,

FGFR mutations lead to gain of function or loss of function. Other molecular alterations in DNET include a single mutation in the promoter of *TERT*, three *IDH1* mutations, gains of chromosomes 5 and 7 as well as 6, loss of heterozygosity (LOH) of 1p/19q or 19q, LOH of 10q (PTEN locus), and BRAF p.V600E mutations (especially DNET in extratemporal locations) [2].

Given that both DNET and ganglioglioma share CD34 expression, *BRAF*[V600E] mutations, and chromosomal copy number profiles with ganglioglioma, they may represent different morphological variants of the same tumor entity. However, while ganglioglioma shows a cellular atypia in both the glial and neuronal components, DNET displays minimal, if any, atypia. Further, whereas ganglioglioma resides subcortically and has a unifocal aspect, DNET typically affects the cortex and has a multinodular aspect.

Occasional multifocal DNET is associated with neurofibromatosis, XXY syndrome, and intradural spinal lipoma.

12.5 Clinical features

Clinically, DNET is associated with perinatal events and familial epilepsy, manifesting as intractable complex partial seizures and frequent secondary generalization (particularly in late childhood, median 7–13 years), as well as infantile spasms. Other clinical symptoms include cognitive impairment (absent or moderate) and relatively frequent mood disorders. The signs of raised intracranial pressure may suggest malignant transformation.

Electroencephalography (EEG) ictal and interictal discharges are predominantly focal or regional in the complex and simple forms, although widespread discharges and discordant abnormalities are observed in the nonspecific form. Some DNET cases may have a normal EEG.

12.6 Diagnosis

Clinical diagnosis of DNET is essentially based on the following criteria: (i) partial seizures, with or without secondary generalization before age 20; (ii) no neurological deficit or presence of a stable and likely congenital neurological deficit; (iii) cortical topography of the lesion on MRI; and (iv) no mass effect on CT or MRI (with the exception of a cyst).

DNET lesion usually has a dimension of 1.0–2.5 cm (rarely up to 7 cm) and appears as a well-defined, solitary nodular mass or poorly demarcated lesion. DNET typically shows a well-demarcated, hypodense cortical lesion with sporadic calcification on CT, and appears as Type 1 (cystic/polycystic-like, well-delineated, strongly hypointense), Type 2 (nodular-like, heterogeneous), or Type 3 (dysplastic-like, isosignal/hyposignal, poor delineated, gray–white matter blurring) on T1-weighted images. Simple or complex DNET is generally Type 1, and nonspecific DNET is either Type 2 or Type 3. DNET is hyperintense on T2-weighted images, with occasional weak enhancement [3,4].

Histologically, DNET is characterized by the specific GN element consisting of small, round monotonous cells (so-called oligodendroglia-like cells or OLC) and floating neurons in an abundant mucinous matrix. The specific GN elements may form typical nodules or show a diffuse pattern. The OLC are arranged in a columnar pattern (of microcystic, alveolar, compact, or targetoid structure) perpendicular to the cortical surface and separated by a mucinous matrix. Floating neurons typically lack perineuronal satellitosis or nuclear atypia. The simple form contains only the specific GN elements without multinodular architecture. The complex form contains the GN elements, glial nodules, and/or FCD of the adjacent cortex. The nonspecific form possesses neither the specific GN elements nor the multinodular architecture but has the clinical and radiological features of a complex DNET. Occasionally, DNET may display increased cellularity, pleomorphism, cytological atypia, microvascular proliferation, calcification, necrosis, high MIB-1-labeling indices, and extensive involvement of surrounding structures [1].

Immunohistochemically, floating neurons are positive for several neuronal markers (e.g., synaptophysin neurofilament, NeuN, neuron-specific enolase, MAP2, and class III beta-tubulin); the OLC is strongly positive for S-100 protein and Oligo-2 but negative for glial fibrillary acidic protein. The nonspecific type of DNET may be weakly positive for synaptophysin but negative for other neuronal markers except MAP2. In addition, the specific GN elements contain CD34-positive cells in a focal pattern, whereas most nonspecific tumors and glial nodules of some complex DNET show CD34-positive cells in focal, multifocal, or diffuse patterns. Therefore, combined analysis of CD34 and MAP2 may be used for differential diagnosis between nonspecific DNET and simple/complex DNET as well as between DNET and epilepsy-associated tumors that mimic DNET [1].

Differential diagnoses of DNET include several long-term epilepsy-associated tumors, such as ganglioglioma, pleomorphic xanthoastrocytoma, papillary glioneuronal tumor, supratentorial pilocytic astrocytoma, diffuse astrocytoma, oligodendroglioma, angiocentric glioma, and extraventricular neurocytoma [5–7].

Ganglioglioma is characterized histologically by the presence of voluminous ganglion-like neurons, granular bodies, perivascular lymphocytic cuffing (often extending within arachnoid spaces), and intercellular reticulinic frame, which are all absent in DNET. Although both ganglioglioma and nonspecific DNET display similar immunoreactivity patterns for CD34, they react quite differently with MAP2 (being faintly positive or negative in ganglioglioma and consistently positive in DNET).

Low-grade diffuse glioma and the nonspecific DNET all contain blurred cortex–white matter interface. However, on T1-weighted images, diffuse glioma appears hypointense, and the nonspecific DNET is isointense to normal cortex. Also, diffuse glioma often harbors IDH1 mutations, while DNET rarely does so.

Additionally, astrocytoma and oligodendroglioma are diffusely positive for MAP2 but negative for CD34 in contrast to DNET that are consistently positive for MAP2 and focally, multifocally, or diffusely positive for CD34. Furthermore, BRAFV600E mutations are absent in oligodendroglioma but present in DNET.

12.7 Treatment

The recommended treatment for DNET is complete surgical resection (e.g., lesionectomy or corticectomy) without adjuvant radiotherapy and chemotherapy, given that radiation therapy is implicated in malignant transformation and that DNET is often resistant to anti-epileptic therapy [8]. Therefore, accurate identification and diagnosis will spare the patients from unnecessary surgery that may lead to potential neurological and cognitive damage. Patients with unresected or incompletely removed DNET or 3 years after complete resection are prone to potential development of pilocytic astrocytoma, recurrence and malignant transformation; long-term follow-up is important [8–10].

12.8 Prognosis

Prognosis of DNET is related to complete tumor resection, young age at surgery, short epilepsy duration, and absence of cortico–subcortical damage. Complete surgical removal of tumor and epileptogenic zones is effective at controlling seizures in over 98% of patients and achieving long-term seizure freedom in 86% of patients. Incomplete resection represents the main cause of surgical failure, and repeated surgical operation is necessary for seizure-free outcome. The rate of seizure freedom 12 months after incomplete resection and gross total resection stands at 52% and 99%, respectively.

References

1. Suh YL. Dysembryoplastic neuroepithelial tumors. *J Pathol Transl Med*. 2015;49(6):438–4.
2. Rivera B, Gayden T, Carrot-Zhang J, et al. Germline and somatic FGFR1 abnormalities in dysembryoplastic neuroepithelial tumors. *Acta Neuropathol*. 2016;131(6):847–63.
3. Paudel K, Borofsky S, Jones RV, Levy LM. Dysembryoplastic neuro-epithelial tumor with atypical presentation: MRI and diffusion tensor characteristics. *J Radiol Case Rep*. 2013;7(11):7–14.
4. Chassoux F, Daumas-Duport C. Dysembryoplastic neuroepithelial tumors: Where are we now? *Epilepsia*. 2013;54 Suppl 9:129–34.
5. Giulioni M, Marucci G, Martinoni M, et al. Epilepsy associated tumors: Review article. *World J Clin Cases*. 2014;2(11):623–41.
6. Keser H, Barnes M, Moes G, Lee HS, Tihan T. Well-differentiated pediatric glial neoplasms with features of oligodendroglioma, angiocentric glioma and dysembryoplastic neuroepithelial tumors: A morphological diagnostic challenge. *Turk Patoloji Derg*. 2014;30(1):23–9.
7. Sukheeja D, Mehta J. Dysembryoplastic neuroepithelial tumor: A rare brain tumor not to be misdiagnosed. *Asian J Neurosurg*. 2016;11(2):174.
8. Ranger A, Diosy D. Seizures in children with dysembryoplastic neuro-epithelial tumors of the brain—A review of surgical outcomes across several studies. *Childs Nerv Syst*. 2015;31(6):847–55.
9. Xu J, Du J, Shan Y. Manifestation and treatment of intraventricular dysembryoplastic neuroepithelial tumor. *Chin Med J (Engl)*. 2014;127(7):1390.
10. Schijns OE, Beckervordersandforth J, Wagner L, Hoogland G. Long-term drug-resistant temporal lobe epilepsy associated with a mixed ganglioglioma and dysembryoplastic neuroepithelial tumor in an elderly patient. *Surg Neurol Int*. 2016;7(Suppl 9):S243–6.

13
Gangliocytoma and Lhermitte–Duclos Disease

13.1 Definition

Within the neuronal and mixed neuronal-glial tumors group, several are known to affect neurons of the central nervous system (CNS). These tumors (also called *ganglion cell tumors*) consist of gangliocytoma, Lhermitte–Duclos disease (LDD, dysplastic cerebellar gangliocytoma), ganglioglioma, and desmoplastic infantile ganglioglioma [1]. Although gangliocytoma resembles ganglioglioma in the presence of neoplastic ganglion cells (large mature neurons with cytological or architectural abnormalities), gangliocytoma differs from ganglioglioma (see Chapter 14) by the absence of neoplastic glial cells [1].

Gangliocytoma is a rare, slow-growing tumor (WHO Grade I) that occurs in the cerebrum (particularly the temporal lobe) as well as other locations within the CNS (e.g., the floor of the third ventricle, cerebellum, brainstem, and upper spinal cord). The tumor typically shows neuronal (ganglion) cells in a sparse glial stroma and does not usually become malignant.

LDD (dysplastic cerebellar gangliocytoma) is a rare, benign lesion (WHO Grade I) that is confined to the cerebellum. LLD is characterized by enlarged cerebellar folia with striking "inverted" architecture and replacement of the internal granular layer by dysplastic ganglion cells. Intermediate zones between the normal and abnormal cerebellar tissues display gradual transitions of large dysplastic cells replacing the small granule cells. Dysplastic areas contain rich vascularization and numerous interstitial vacuoles. Associated with Cowden syndrome, LDD demonstrates very low proliferative activity, although progression to malignant and other benign tumors such as anaplastic ganglioglioma and dysembryoplastic neuroepithelial tumor has been reported in a few cases [2,3].

13.2 Biology

Derived from the ectoderm, the neural crest comprises a population of cells (i.e., neural crest cells) that emigrate from the dorsal neural tube during early embryogenesis to various locations and differentiate into a myriad of cell types within the embryo. These include peripheral and enteric neurons and glia, melanocytes, craniofacial cartilage and bone, and smooth muscle.

Neuron is the basic cell of the nervous system that contains a nucleus within a cell body (perikaryon) and extends one or more processes (usually an axon and one or more dendrites). A neuron with an axon only is classified as *unipolar neuron*, that with an axon and a dendrite is classified as *bipolar neuron*, and that with an axon and two or more dendrites is classified as *multipolar neuron*, which is the most common type and widely distributed in the CNS. The axon conducts the impulses to the dendrite of another neuron or to an effector organ. The dendrites receive stimuli from a receptor organ or other nerves and transmit through the neuron to the axon. According to the direction in which they conduct impulses, neurons are categorized into three groups: (i) *afferent* or *sensory neurons* (which conduct impulses from a receptor to a center), (ii) *efferent* or *motor neurons* (which carry impulses away from a center to an organ of response), and (iii) *interneurons* (which conduct impulses from afferent to efferent neurons). The point at which an impulse is transmitted from one neuron to another is known as synapse.

Ganglion cell once used to refer to any neuron is now more commonly known as a neuron whose cell body is located outside the limits of the brain and spinal cord, thus forming part of the peripheral nervous system. Ganglion cell may be either the pseudounipolar cell of the sensory spinal and cranial nerves (sensory ganglia) or the peripheral multipolar motor neuron innervating the viscera (visceral or autonomic ganglia).

Gangliocytoma evolves from neural crest cells in the temporal lobe (of the cerebrum) and the floor of the third ventricle, in addition to the cerebellum, parieto-occipital region, frontal lobe, brainstem, and spinal cord. The intramedullary form of spinal gangliocytoma (involving the thoracolumbar region and the cervical spine through the neck area) accounts for <10% of all gangliocytoma cases [4].

LDD (synonyms: dysplastic gangliocytoma of the cerebellum, dysplastic gangliocytoma, purkinjeoma, ganglioneuroma, granular or granulomolecular hypertrophy of the cerebellum, diffuse hypertrophy of the cerebellar cortex, gangliomatosis of the cerebellum, hamartoma of the cerebellum, myelinated neurocytoma, and gangliocytoma myelinicum diffusum) occurs only in the cerebellum (especially the left cerebellar hemisphere and rarely extending into the vermis, which is the medial, cortico-nuclear zone of the cerebellum). LDD contains abnormal hypertrophic ganglion cells somewhat similar to Purkinje cells, leading to structural alteration in the cerebellar cortex and reduction of white matter in the cerebellum [5].

13.3 Epidemiology

Representing 0.1%–0.5% of CNS tumors, gangliocytoma occurs at the extracranial and intracranial sites of all age groups, with 40%–60% of cases involving adolescents and young adults (of 10–30 years old, with a mean age of 25 years).

Lhermitte–Duclos disease is a rare entity, with about 200 cases reported to date. Although onset of LDD may occur at any age, it often becomes clinically apparent in the third and fourth decades [6].

13.4 Pathogenesis

Patients with recurrent gangliocytoma are found to harbor mutation in the TP53 gene but not in the epidermal growth factor receptor (EGFR) gene.

Adult-onset Lhermitte–Duclos disease (LDD) appears to be a pathognomonic manifestation of Cowden disease. Molecular analysis indicates that Cowden disease is linked to germline mutations (80%) and mutations in promoter region in the phosphatase and tensin homolog (PTEN) gene located on chromosome 10q22–23. The *PTEN* gene (also known as *MMAC1* for *mutated in multiple advanced cancers*, or TEP1 for *TGF-β-regulated and epithelial cell–enriched phosphatase*) comprises nine exons and encodes a 403 amino-acid lipid phosphatase for the phosphatidylinositol 3-kinase pathway. The PTEN protein dephosphorylates PIP_3 and PIP_2, inhibits formation of phosphorylated AKT (a serine/threonine kinase), and increases apoptosis. A loss or reduction in PTEN activity leads to increased activity of the AKT and mTOR pathways and unrestricted cell proliferation, hypertrophy, and improper migration, with clinical symptoms ranging from macrocephaly, facial trichilemmomas, acral keratoses, and papillomatous papules to breast, thyroid, and endometrial carcinomas. A genetic disorder of autosomal dominant inheritance, Cowden disease is familial in half of all cases and spontaneous in the other half. Most LDD patients have a germline loss of one *PTEN* allele and suffer from loss of the remaining *PTEN* allele at some point, leading to abnormal growth of the granule cells. In addition to Cowden disease (and LDD), *PTEN*-related hamartomatous tumor syndromes include Bannayan–Riley–Ruvalcaba, Proteus, and Proteus-like syndromes. However, *PTEN* mutations have not been identified in childhood onset LDD cases, suggestive of spontaneous occurrence [2,3].

13.5 Clinical features

Depending on the tumor's location, clinical symptoms of gangliocytoma may range from seizures, increased brain pressure, endocrine disorders, focal symptoms, anxious/depressed mood, insomnia, fatigue, and pain to no symptom [1].

Patients with spinal gangliocytoma may develop radiculopathy (a condition of the nerve roots), paraparesis (partial paralysis of the legs), or cauda equina syndrome. Those with tumors in the cerebral cortex often show epilepsy, and those with thoracic dumbbell gangliocytomas have scoliosis as a presenting sign [4].

Clinical symptoms of LDD include intermittent headache, nausea, cranial nerve palsies, gait abnormality, ataxia (problem with movement and coordination), tremor, visual disturbances, abnormal EEG, diplopia, megalencephaly (enlarged brain), polydactylia (extra fingers or toes), macroglossia (large tongue), subcortical gray matter heterotopia, meningeal glial heterotopia, vascular malformations, and skull abnormalities. Most of these symptoms may be attributable to raised intracranial pressure, obstructive hydrocephalus, and cerebellar dysfunction related to the growing tumor mass [6].

13.6 Diagnosis

Gangliocytoma is a cortical solid lesion with little mass effect and minimal surrounding vasogenic edema. On MRI, gangliocytoma is a circumscribed solid or mixed solid and cystic mass spanning a long segment of the cord, showing hypointense T1 signal and hyperintense T2 signal in cystic areas (with calcification appearing hypointense). Enhancement patterns range from minimal to marked and may be solid, rim, or nodular. In addition to reactive scoliosis, adjacent cord edema, syringomyelia, and peritumoral cysts may be present.

Histologically, gangliocytoma demonstrates abnormal mature ganglion cells and finely fibrillar neuropil-rich stroma with notable absence of glial cells. It is negative for glial fibrillary acidic protein.

Differential diagnosis of ganglioglioma includes astrocytoma, ependymoma, hemangioblastoma, and paraganglioma. It is notable that astrocytoma has poorly defined margins, whereas ependymoma shows a central location in the spinal cord and hemorrhage. Hemangioblastoma and paraganglioma are unusual intramedullary tumors.

Lhermitte–Duclos disease lesion is a non-neoplastic (and probably hamartomatous) mass showing thickening of the outer molecular cell layer, loss of the middle Purkinje cell layer, and infiltration of the inner granular cell layer with dysplastic ganglion cells. On CT, the tumor is a well-defined lesion mixed with an area of calcification in the right cerebellum and obstructive hydrocephalus. On MRI, the tumor reveals characteristic nonenhancing gyriform (striated/tigroid) patterns with enlargement of the cerebellar folia. It appears hypointense on T1 and hyperintense with preserved cortical striations on T2, with hyperintensity due to the T2 shine-through effect on diffusion-weighted imaging (DWI) and rare superficial enhancement due possibly to vascular proliferation on T1 C+. MR spectroscopy shows elevated lactate, slightly reduced *N-acetyl aspartate (NAA)* (by about 10%), reduced myo-inositol (by 30%–80%), reduced choline (by 20%–50%), and reduced Cho/Cr ratio [7,8].

Microscopically, LDD lesions display massive replacement and expansion of the internal granule cell layer by large hypertrophic neurons with vesicular nuclei and prominent nucleoli. The outer molecular layer is widened by the abundant, enlarged, irregularly myelinated axons from hypertrophic granule cells. Clear vacuoles are found in white matter and molecular layer. No mitosis, necrosis, and endothelial proliferation are observed. Calcification and ectatic vessels are common. Unlike glioma with diffuse infiltration, LDD lesion shows larger, more rounded, more uniform neurons in clusters. Dysplastic ganglion cells in LDD are positive for synaptophysin, phosphorylated AKT, and phosphorylated S6 (indicates AKT/mTOR pathway activation), with loss of PTEN protein expression. Differential diagnosis of LDD consists of ganglion cell tumor, infiltrating glioma with entrapped neurons, and cerebellitis [5,8].

13.7 Treatment

Gangliocytoma is a low-grade tumor, for which surgery alone is often sufficient. Complete resection of supratentorial gangliocytomas is achievable in >75% of cases, and clinically relevant recurrence/regrowth of the tumor is rare even after partial resection [9]. Treatment for symptomatic LDD patients includes surgical debulking of the tumor, if complete resection is not possible [3,10].

13.8 Prognosis

Gangliocytoma is a benign tumor without undergoing anaplastic change and is usually associated with long-term survival. Patients with resected spinal cord ganglioglioma have 5- and 10-year survival rates of 89% and 83%, respectively.

LDD has a good prognosis, with recurrence in 25% of cases, and rare malignant transformation.

References

1. Adesina AM, Rauch RA. Ganglioglioma and Gangliocytoma. In: Adesina AM, Tihan T, Fuller CE, Poussaint TY. (eds), *Atlas of Pediatric Brain Tumors.* Springer, 2010, pp. 181–91.
2. Govindan A, Premkumar S, Alapatt JP. Lhermitte-Duclos disease (dysplastic gangliocytoma of the cerebellum) as a component of Cowden syndrome. *Indian J Pathol Microbiol.* 2012;55(1):107–8.
3. Golden N, Tjokorda MG, Sri M, Niryana W, Herman S. Management of unusual dysplastic gangliocytoma of the cerebellum (Lhermitte-Duclos disease) in a developing country: Case report and review of the literature. *Asian J Neurosurg.* 2016;11(2):170.
4. Oppenheimer DC, Johnson MD, Judkins AR. Ganglioglioma of the spinal cord. *J Clin Imaging Sci.* 2015;5:53.
5. Rusiecki D, Lach B. Lhermitte-Duclos disease with neurofibrillary tangles in heterotopic cerebral grey matter. *Folia Neuropathol.* 2016;54(2):190–6.
6. Assarzadegan F, Gharib A, Behbahani S, Ebrahimi-Abyaneh M. Intracranial hypertension and cerebellar symptoms due to Lhermitte-Duclos disease. *Iran J Neurol.* 2015;14(2):113–15.
7. Calabria F, Grillea G, Zinzi M, et al. Lhermitte-Duclos disease presenting with positron emission tomography-magnetic resonance fusion imaging: A case report. *J Med Case Rep.* 2012;6:76.
8. Radiopedia.org. Gangliocytomas. https://radiopaedia.org/articles/gangliocytoma. Accessed November 7, 2016.
9. Furie DM, Felsberg GJ, Tien RD, et al. MRI of gangliocytoma of the cerebellum and spinal cord. *J Comput Assist Tomogr.* 1993;17(3):488–91.
10. Ozeren E, Gurses L, Sorar M, Er U, Önder E, Arıkök AT. L'hermitte-Duclos disease in an elderly patient: A case report and review of the literature. *Asian J Neurosurg.* 2014;9(4):246.

14

Ganglioglioma

Mandana Behbahani, Hasan R. Syed,
Tadanori Tomita, and Christopher Kalhorn

14.1 Definition

Ganglioglioma (GG) is a well-differentiated, slow-growing, neuroepithelial tumor (WHO Grade I), consisting of neoplastic mature ganglionic and glial cells. This tumor often occurs in the cerebral hemispheres and is associated with chronic epilepsy in patients.

A rare and malignant variant of ganglioglioma is referred to as *anaplastic ganglioglioma* (WHO Grade III). This variant shows anaplastic changes (e.g., increased mitotic activity, vascular proliferation, and necrosis) in the glial component but seldom in the neuronal component, resulting in unusual aggressiveness and poor outcome [1].

14.2 Biology

Based on the combination of slow evolution, the extended period of symptomatology prior to diagnosis, and multiple histopathological studies, ganglioglioma was postulated to have a hamartomatous or maldevelopmental origin. According to the maldevelopmental theory of origin, ganglioglioma may derive from foci of cortical dysplasia, given its location and epileptogenic presentation. Ectopic neuronal cell rests derived from peripheral autonomic nervous tissue or possibly the presence of a single stem cell with the ability to differentiate along both glial and neuronal cell lines may give rise to ganglioglioma. Clonality studies have indicated the transformation of a common putative neuroglial precursor cell to ganglioglioma. More recently, molecular alteration involving the phosphatidylinositol 3 kinase (PI3K)–mechanistic target of rapamycin (mTOR) pathway has been implicated in the pathogenesis of glioneuronal lesions, contributing to the maldevelopment of these tumors. Overall, the distinct histopathological characteristics of ganglioglioma, the coexistence of cortical dysplasia, and the expression of stem cell markers (e.g., CD34) suggest a developmental origin for these lesions [2].

Although considered benign, ganglioglioma has the potential to undergo malignant transformation in the glial component of the tumor and on occasion has been reported to metastasize [3,4]. The tumor cells display markers consistent with a slow-growing benign tumor, including low levels of Ki-67 and MIB1 and EGFR amplification, without p53 mutation. However, in malignant recurrence, p53 mutation and loss of P6 expression are observable with elevated levels of EGFR.

An epileptogenic tumor, ganglioglioma is identified as the structural lesion associated with chronic temporal lobe epilepsy in 20%–40% of patients undergoing surgical intervention for chronic, refractory epilepsy.

Ganglioglioma has a predilection for the supratentorial region and particularly that of mesiotemporal locations, in up to 70% of cases [1]. Despite its common occurrence in the mesiotemporal region, ganglioglioma has also been reported to occur less frequently in the infratentorial fossa, involving the cerebellum, as well as within the spinal cord, optic nerve, pituitary, and pineal glands [1].

14.3 Epidemiology

Ganglioglioma is rare, accounting for only 0.4%–1.3% of all CNS neoplasms but about 70% of neuronal and mixed neuronal-glial tumors [1]. The tumor most commonly affects children (3.5% in children of <1 year) and youth (63% in children of >10 years) and occasionally people of other ages. There is a slight predilection for males relative to female (59.8% male vs. 40.2% female—a ratio of 1.5:1) [1,5].

14.4 Pathogenesis

Ganglioglioma is associated with the gains of chromosome 7 (21%), as well as chromosomes 5, 8, and 12, in addition to the loss of CDKN2A/B and DMBT1 or a gain/amplification of CDK4 [6]. Interestingly, ganglioglioma in patients with long-standing epilepsy contains a lower number of genetic abnormalities compared to that in patients without long-standing epilepsy; this finding lends support to the observation that ganglioglioma patients with long-standing epilepsy have a lower recurrence rate and a better clinical course than those without epilepsy [2].

Most recently, the PI3K-mTOR pathway, which is responsible for cell size, growth control, cortical development, and neuronal migration, has been shown to play a critical role in the specific pathogenesis of gangliogioma [7,8].

Mutational analysis of downstream tumor suppressor complexes involving TSC1 (hamartin) and TSC2 (tuberin) reveals gene alteration in TSC2, including polymorphism in intron 4 and exon 41 to be overrepresented in ganglioglioma. Concurrently, somatic mutation in intron 32 is identified in the glial portion but not within the neurons of ganglioglioma. An increased polymorphism within tuberin has been noted in ganglioglioma relative to that of normal brain tissue. In contrast, ezrin–radixin–moesin (ERB) proteins interacting with TSC1 to regulate cell adhesion and migration display high levels of aberrancy within dysplastic elements of glioneuronal lesions, such as ganglioglioma. Additionally, LIM domain-binding 2, a gene known to play a role in brain development, is reduced in expression, possibly highlighting the development of an aberrant neuronal network as a major etiology in ganglioglioma, as previously hypothesized [9]. Interestingly, recent genetic studies suggest against the involvement of known pathogenic genes, such as TP53, EGFR, and PTN, as contributors to gangliogioma [8].

14.5 Clinical features

The clinical presentation of patients with ganglioglioma inevitably depends on the tumor location and size, with symptoms ranging from headache, vomiting, nausea, weakness, unsteadiness, incoordination, drowsiness, increased intracranial pressure, abnormal sensations, seizures, involuntary muscle movements, abnormal eye movements, impaired vision, impaired memory, and personality changes to cognitive, behavioral, balance, and speech problems.

14.6 Diagnosis

Computed tomography (CT) of ganglioglioma demonstrates a low-density, well-circumscribed lesion. Often calcifications can be identified within the solid portion of the lesion. The solid portion itself can appear as isodense, hypodense, or mixed. There is a variable degree of contrast enhancement noted on CT; however, enhancement is generally scant or absent.

Magnetic resonance imaging (MRI) shows a variable T1 signal intensity ranging from most often isointense to generally hypointense, whereas T2-weighted sequences reveal a hyperintense signal. On MRI, these tumors can appear a solid, cystic, or cystic with mural nodules. On MRI, similar to CT, contrast enhancement within the solid portion of the tumor is often variable, generally with little to no enhancement.

Gross pathology reveals a glossy, grayish-yellow, solid lesion, which is often associated with a cystic nodular component. There can be focal regions of mineralization on exploration of the surgical specimen, correlating to regions of calcification on imaging. Hemorrhage and necrosis are rarely encountered, unless there is malignant transformation [1].

Histopathologically, ganglioglioma contains hallmarks of a heterogeneous mix of neuronal and glial components. Neurons visualized are often multipolar with dysplastic features within a non-neoplastic, glial stroma and network of reticulin fibers. Dysplastic neurons are characterized as loss of cyto-architectural organization, abnormal localization, clustered appearance, cytomegaly, perimembranous aggregated Nissl substance, or presence of bi- or multinucleated neurons. Within the neuronal component, there is vast variability that can resemble the range from fibrillary astrocytoma to oligodendroglioma and pilocytic astrocytoma. There are often characteristic features such as calcification, lymphoid infiltration of perivascular space, a prominent capillary network, and a microcystic component, without any gross signs of hemorrhage or necrosis. In cases of malignant transformation, necrosis and hemorrhage are not unexpected findings. Similarly, mitotic figures have been observed, though rare, and correlative Ki-67/MIB-1 labeling of neuronal component yields low range of 1.1%–2.7%, unless there is malignant transformation [1].

Immunohistochemically, MAP2 reactivity is absent in astrocytic components of ganglioglioma, whereas glial fibrillary acidic protein (GFAP) staining demonstrates the astrocytic component that usually gives rise to the neoplastic glial element of ganglioglioma. Overall, ganglioglioma is characteristic of CD34 detection (a stem cell epitope), lack of glial MAP2 staining, low Ki-67, variable and nonspecific GFAP staining, and no p53 immunoreactivity [1,7].

14.7 Treatment

Surgical intervention with the goal of gross total resection is the first modality of treatment for ganglioglioma. Early surgical resection of ganglioglioma reduces both morbidity and mortality as it pertains to seizures as well as recurrence. Seizure freedom postoperatively ranges from 63% to 90%. Despite the capability to obtain seizure control with partial resection, the literature advocates for complete resection for optimal prognosis [1,10]. In case of WHO Grade II or III, which indicates a higher likelihood of recurrence, long-term clinical follow-up with serial imaging should be offered.

14.8 Prognosis

The calculated 7.5-year survival rate for ganglioglioma patients is 80%, with a 7.5-year recurrence-free survival rate of 97%. Tumor recurrence is noted in only 1% of the patients with WHO Grade I lesion, compared to 18% of patients with Grade II lesion and 50% of patients with Grade III lesion. Only 1.6% of patients develop malignant transformation.

Features associated with poor prognosis are older age at time of surgery, subtotal resection, extratemporal location, and absence of chronic epilepsy. Malignant recurrence or transformation of ganglioglioma is seen in tumors with elevated cellularity and mitotic activity, anaplastic glial features, microvascular proliferation, necrosis, presence of gemistocytic component, and CD34 focal tumor labeling [7].

References

1. Louis DN, Perry A, Reifenberger G, et al. The 2016 World Health Organization classification of tumors of the central nervous system: A summary. *Acta Neuropathol.* 2016; 131: 803–20.
2. Aronica E, Boer K, Becker A, et al. Gene expression profile analysis of epilepsy-associated gangliogliomas. *Neuroscience.* 2008; 151(1): 272–92.
3. Araki M, Fan J, Haraoka S, Moritake T, Yoshii Y, Watanabe T. Extracranial metastasis of anaplastic ganglioglioma through a ventriculoperitoneal shunt: A case report. *Pathol Int.* 1999; 49(3): 258–3.
4. Jay V, Squire J, Blaser S, Hoffman HJ, Hwang P. Intracranial and spinal metastases from a ganglioglioma with unusual cytogenetic abnormalities in a patient with complex partial seizures. *Childs Nerv Syst.* 1997; 13(10): 550–5.
5. Dudley RW, Torok MR, Gallegos DR, et al. Pediatric low-grade ganglioglioma: Epidemiology, treatments, and outcome analysis on 348 children from the surveillance, epidemiology, and end results database. *Neurosurgery.* 2015; 76(3): 313–19; discussion 319; quiz 319–20.
6. Hoischen A, Ehrler M, Fassunke J, et al. Comprehensive characterization of genomic aberrations in gangliogliomas by CGH, array-based CGH and interphase FISH. *Brain Pathol.* 2008; 18(3): 326–37.
7. Aronica E, Niehusman P. Gangliogliomas: Molecular pathogenesis and epileptogenesis. In: Hayat MA. (ed.) *Tumors of the Central Nervous System. Vol 5. Astrocytomas, Hemangioblastomas and Gangliogliomas.* Springer, the Netherlands, 2011, pp. 253–65.

8. Becker AJ, Blumcke I, Urbach H, Hans V, Majores M. Molecular neuropathology of epilepsy-associated glioneuronal malformations. *J Neuropathol Exp Neurol.* 2006; 65(2): 99–108.

9. Fassunke J, Majores M, Tresch A, et al. Array analysis of epilepsy-associated gangliogliomas reveals expression patterns related to aberrant development of neuronal precursors. *Brain.* 2008; 131(11): 3034–50.

10. Giulioni M, Gardella E, Rubboli G, et al. Lesionectomy in epileptogenic gangliogliomas: Seizure outcome and surgical results. *J Clin Neurosci.* 2006; 13(5): 529–35.

15
Neurocytoma

15.1 Definition

Neurocytoma (commonly known as *central neurocytoma* or *CN*) is a neuronal and mixed neuronal–glial neoplasm that typically arises from the neuronal cells of the septum pellucidum, the third ventricle, and the lateral ventricles. Neurocytoma is a WHO Grade II tumor with an MIB-1 labeling index (MIB-1 LI) of <2% that may grow inwards into the ventricular system, forming interventricular neurocytoma and leading to blurred vision and increased intracranial pressure. In addition, there exist some "atypical CNs" with unusual aggressiveness, which demonstrate elevated MIB-1 LI >2% and/or histological atypical features, with increased tendency to spread through the cerebrospinal fluid, causing craniospinal axis dissemination [1,2].

Occasionally, neurocytoma may occur outside of the ventricle and become extraventricular neurocytoma (EVN). Often located at the frontal lobe and parietal lobe, as well as the cerebellum, thalamus, brainstem, sellar and spinal cord, retroperitoneum, abdomen, and pelvis, EVN exhibits varied morphological characteristics, cellularity, and proliferation rate with aggressive histological features. Collectively, CN and EVN are sometimes referred to as *cerebral neurocytomas* [2].

Besides cerebral neurocytomas, there is another brain tumor known as *cerebellar liponeurocytoma* (CLPN, WHO Grade II), which is defined as a rare, well-differentiated neurocytic tumor of the cerebellum with focal or regional lipomatous differentiation [3,4].

15.2 Biology

Neurocytoma (CN, alternatively known as *neuroepithelioma*) is a tumor derived primarily from bipotential precursor cells in the fornix or walls of the lateral ventricles or septum pellucidum in the region of the Monro foramina. It is mainly located in the midline supratentorially, more commonly on the right side. Most CN (about 50%) are found in the anterior portion of the lateral ventricle, followed by the lateral and third ventricles (15%), biventricular location (13%), and the third ventricle alone (3%).

EVN is located in the occipital lobe, parietal lobe, temporal lobe, frontal lobe, hypothalamus, cerebellum, pons, spinal cord, cauda equina, and retina, with similar behavior to its intraventricular counterparts. Neurocytoma originated from the sellar region is called *extraventricular neurocytoma of the sellar region* (EVNSR), which may be misdiagnosed as other diseases (e.g., pituitary tumor, craniopharyngioma, and meningioma) [5].

CLPN (formerly lipomatous medulloblastoma, neurolipocytoma, medullocytoma, and lipomatous glioneurocytoma) arises from the neurocytes admixed with lipidized cells (due to tumoral lipidization instead of adipose metaplasia) in the cerebellum (86%, usually hemispheric, less commonly vermian or cerebellopontine angle) and occasionally in the supratentorial compartment (14%) or the fourth ventricle. The tumor is characterized by many lipidized cells found in clusters or scattered between small neoplastic cells, with both neuronal and glial differentiation [4].

15.3 Epidemiology

Neurocytoma is a rare, mostly benign neoplasm of the CNS, accounting for only 0.1%–0.5% of all intracranial tumors. Typically affecting young adults around the third decade (ranging from 2 to 70 years), it constitutes nearly 50% of adult supratentorial intraventricular tumors, with a higher prevalence in Asian populations than in Caucasians.

EVN represents 0.02%–0.10% of intracranial tumors and affects patients of 5–76 years of age (average 34 years), with a high incidence among Asians, especially the Japanese. EVNSR occurs in patients of 25–66 years of age (mean age 45 years) with a male-to-female ratio of 1:2.

CLPN affects patients aged 4–69 years (median 49 years), with a slight female predominance (1: 1.8).

15.4 Pathogenesis

Neurocytoma (CN) often contains loss of heterozygosity of 1p and 19q13.2–13.4. EVN is associated with deletion of chromosome arms 1p/19q either in isolation or in combination and with translocation t(1;19). In fact, 1p19q loss and t(1;19) may be linked to aggressive histology in EVN. CLPN harbors TP53 missense mutations and demonstrates overexpression of NEUROG1 and FABP4.

15.5 Clinical features

Neurocytoma (CN) is typically associated with increased intracranial pressure secondary to obstructive hydrocephalus, leading to headache (93%), visual changes (37%), nausea and vomiting (30%), lightheadedness, impaired mental activity, and gait instability. In rare and extreme cases, memory disturbance, dementia, hemiparesis, seizures, hemorrhage, and coma may be observed [6].

EVN may cause headache, nausea, vomiting, anxiety, hydrocephalus, chronic intracranial hypertension, spinal pain, somnolence, hemiparesis (weakness of the entire left or right side), paraparesis (partial paralysis of the lower limbs), paresthesia (pins and needles), bilateral papilloedema (vision changes), urinary incontinence, speech and learning delay, attention deficit, and seizures. EVNSR may manifest as hypopsia, vision field defection, headache, dizziness, and hypertensive intracranial syndrome [2,7].

CLPN may cause headache, vomiting, vertigo, hemiparesis, altered consciousness, dizziness, unsteadiness, gait disturbance, frequent falls, visual changes, and increased intracranial pressure [4].

15.6 Diagnosis

Diagnosis of CN, EVN, and CLPN is based on a combination of clinical manifestations, radiographic features, and histopathological findings.

Neurocytoma (CN) is usually a soft, tan-colored, well-demarcated mass with attachment to the surrounding ventricular surfaces. On CT, CN is isoattenuating or slightly hyperattenuating, showing a well-demarcated, lobulated mass with moderate to strong heterogeneous contrast enhancement. Clumped, amorphous, or globular calcification is observed in about 50% of cases; cystic changes may be also present. On MRI, CN is heterogeneous on all sequences, but it appears isointense to the cerebral cortex on T1-weighted images and isointense to hyperintense on T2-weighted images. Areas of low signal intensity or absent signal on both T1- and T2-weighted images may be indicative of calcification, cyst, hemorrhage, and tumor vessels.

Histologically, CN is characterized by small, monotonous, round tumor cells with neuronal differentiation that have scant cytoplasm (empty "halo" or "fried egg" appearance), oval nuclei with fine granular chromatin ("salt and pepper" appearance), and inconspicuous nucleoli in the background of the fibrillary matrix and reactive astrocytes. Atypical CN with elevated MIB-1 LI of more than 2%–3% and/or associated histological atypia may show infiltrative margins, increased mitotic activity, cellular pleomorphism, endothelial proliferation, presence of necrosis, or vascular proliferation. CN is positive for markers of neuronal differentiation: synaptophysin (SYN, the most reliable marker, 86%), neuron-specific enolase (NSE), neuronal nuclear antigen, S-100 protein, Leu-7, and focal glial fibrillary acidic protein (GFAP, indicative of glial differentiation of bipotential-astrocytic and neuronal-precursor cells, and correlating with a more malignant disease course). However, CN is negative for vimentin, oligodendrocyte transcription factor and p53 immunoexpression [2,8].

EVN is a discrete, sometimes large, complex, and variably enhancing mass with cystic components (up to 50%). On T1-weighted MRI images, the signal of solid mass is isointense or slightly hypointense, and on T2 hyperintense images there is enhancement after gadolinium administration. Proton MR spectroscopy reveals elevated choline, decreased N-acetylaspartate, and decreased creatine. EVN is positive for SYN and NSE but negative for vimentin, GFAP, S-100, nestin and epithelial membrane antigen (EMA), as well as p53 immunoexpression.

CLPN is a well-circumscribed tumor of cerebellum, which on T1-weighted images is iso- or hypointense with heterogeneous contrast enhancement. On T2-weighted images, the solid component is slightly hyperintense with focal pronounced hyperintense areas corresponding to fat. Some cystic components may be present. Histologically, CLPN contains small, round to ovoid neurocytes arranged in densely cellular sheets with scanty and often clear cytoplasm, rounded to oval nuclei, and salt and pepper chromatin. Lipidized cells containing nonmembrane-bound lipid (resembling mature adipocytes) are interspersed throughout the lesion. Areas of neuronal differentiation are positive for SYN, NSE, MAP-2, and GFAP.

Differential diagnoses of CN, EVN, and CLPN include clear cell ependymoma (rare in posterior fossa, perivascular, negative for SYN, and positive for vimentin), medulloblastoma (common in children, atypia, high proliferation index, and no lipidized cells), oligodendroglioma (rare in posterior fossa, no neurocytic component, no lipidized cells, negative for SYN, and positive for vimentin), and rosette-forming glioneuronal tumor (RGNT) (neurocytic rosettes and no lipidized cells).

15.7 Treatment

Treatment options for CN, EVN, and CLPN consist of surgery (complete or incomplete resection) followed by radiotherapy and chemotherapy. Stereotactic radiosurgery can be an effective and safe alternative for recurrent or residual CN. Postoperative radiotherapy (at a dose of 54–60 Gy) improves the local control and survival rates of EVN patients with incomplete resection. Chemotherapy using carboplatin + VP-16 + ifosfamide or cisplatin + VP-16 + cyclophosphamide is valuable for CN, while that based on procarbazine, lomustine, and vincristine has clear benefits in treating recurrent EVN. Total resection of CLPN gives a good outcome, while subtotal resection of CLPN showing high Ki-67 index is prone to recur. Radiation is thus recommended if there is residual tumor, recurrence, or high Ki-67 index [3].

15.8 Prognosis

Neurocytoma (CN) is a benign tumor with an excellent prognosis. Patients with CN undergoing gross total resection have a 5-year survival rate of up to 99%, compared to 86% for those undergoing subtotal resection. Indeed, subtotal resection appears to be associated with increased risk of recurrence and decreased rate of survival among CN patients. Therefore, radiotherapy and radiosurgery are recommended as an adjuvant treatment to improve survival rate when the tumor is aggressive or only suitable for subtotal resection. The monoclonal antibody MIB-1 LI, which helps distinguish between CN and atypical neurocytoma, provides the best predictor of proliferative potential and clinical outcome of CN. Patients with MIB-1 LI <2% have a relapse rate of 22%, compared to 63% in patients whose MIB-1 LI is >2% [1]. The 5-year recurrence rate in patients with EVN is 44% [9], and recurrence rates of CLPN are estimated at 60% [10]. Recurrent CLPN may show increased mitotic activity, increased proliferative activity (as assessed by MIB-1 staining), vascular proliferation, and necrosis [10].

References

1. Vajrala G, Jain PK, Surana S, Madigubba S, Immaneni SR, Panigrahi MK. Atypical neurocytoma: Dilemma in diagnosis and management. *Surg Neurol Int*. 2014;5:183.
2. Ahmad Z, Din NU, Memon A, Tariq MU, Idrees R, Hasan S. Central, extraventricular and atypical neurocytomas: A clinicopathologic study of 35 cases from Pakistan plus a detailed review of the published literature. *Asian Pac J Cancer Prev*. 2016;17(3):1565–70.
3. Tucker A, Boon-Unge K, McLaughlin N, et al. Cerebellar liponeurocytoma: Relevant clinical cytogenetic findings. *J Pathol Transl Med*. 2016 Oct 16. doi: 10.4132. Epub.
4. Wang KE, Ni M, Wang L, et al. Cerebellar liponeurocytoma: A case report and review of the literature. *Oncol Lett*. 2016;11(2):1061–4.

5. Wang J, Song DL, Deng L, et al. Extraventricular neurocytoma of the sellar region: Case report and literature review. *Springerplus*. 2016; 5(1):987.

6. Montano N, Di Bonaventura R, Coli A, Fernandez E, Meglio M. Primary intramedullary neurocytoma: Case report and literature analysis. *Surg Neurol Int*. 2015;6:178.

7. Song Y, Kang X, Cao G, et al. Clinical characteristics and prognostic factors of brain central neurocytoma. *Oncotarget*. 2016;7(46):76291–7.

8. Jhawar SS, Nadkarni T. Pan ventricular neurocytoma. *Asian J Neurosurg*. 2015;10(1):60.

9. Ji YC, Hu JX, Li Y, Yan PX, Zuo HC. Extraventricular neurocytoma in the left temporal lobe: A case report and review of the literature. *Oncol Lett*. 2016;11(6):3579–82.

10. Nishimoto T, Kaya B. Cerebellar liponeurocytoma. *Arch Pathol Lab Med*. 2012;136(8):965–9.

16
Papillary Glioneuronal Tumor

Andreas K. Demetriades

16.1 Definition

Papillary glioneuronal tumor (PGNT) is a rare cerebral neoplasm, classified by the World Health Organization (WHO) in 2007 as a Grade I neuronal–glial tumor.

First described in 1998 as a new variant of mixed glioneuronal tumor by Komori et al., PGNT is composed primarily of glioneuronal elements with prominent pseudopapillary structures. Whereas the 2000 WHO classification recognized this low-grade tumor as a variant of ganglioglioma, the WHO in 2007 classified it as a Grade I neuronal–glial tumor. This classification was made because of its indolent clinical course and its biphasic neurocytic and glial components arranged in papillary configurations, with the former component in the interpapillary regions and the latter arranged around blood vessels [1]. This unusual tumor can be challenging to the practicing pathologist.

16.2 Biology

PGNT most commonly occurs in the cerebral hemisphere, including the frontal lobe (40%), temporal lobe (30%), parietal lobe (20%), and periventricular site.

16.3 Epidemiology

Although PGNT mainly manifests in young adults, it has occurred in a wide range of ages, including children and the elderly. The mean age at presentation is 27 years, with an age range of 11–75 years [2]. There is a predilection for young adulthood [3], and a very slight predilection for male gender (1.3:1).

16.4 Pathogenesis

The rarity of PGNT and scarcity of tissue for genetic or molecular analysis accounts for the as-yet poor understanding of its biological behavior.

It has been suggested that PGNT involves MAPK signaling pathway deregulation by the presence of *SLC44A1-PRKCA* fusion, which could drive tumorigenesis [4].

16.5 Clinical features

Patients with PGNT often present with headaches (58%), seizures (40%), nausea/vomiting (16%), visual disturbance (10%), speech disturbance (6%), hemiparesis (4%), loss of consciousness (4%), vertigo/ataxia (3%), or no symptoms (10%).

16.6 Diagnosis

Radiology: MRI shows a solid mass, with or without a cystic component, usually involving the cerebral parenchyma.

Histology consists of moderate cellularity without necrosis, microvascular proliferation, or mitotic figures. There is a pseudopapillary pattern of architecture with a single or pseudostratified layer of glial cells overlying hyalinized vessels with interpapillary regions of neurocytic or ganglion cells. Hemosiderin, Rosenthal fibers, peripheral eosinophilic granular bodies, and calcification have also been noted [1,5]. The salient features are summarized in Table 16.1.

Immunohistochemistry includes glial fibrillary acidic protein positivity in the glial cells adjacent to the papillary core and synaptophysin-positive staining in the neuronal component, which also stains for other neuronal antigens such NSE, NeuN, and class III β-tubulin. MIB-1 labelling is in the range of 1%–2%.

Fluorescence *in situ* hybridization studies did not detect 1p deletion. Loss of heterozygosity analysis has also not shown a deletion of either chromosome 1p or 19q.

The proximity of many tumors to the lateral ventricles raises the possibility of origin from multipotential subependymal stem cells. Evidence for possible neuroepithelial stem cell origin, with biphenotypic differentiation, also comes from Govindan et al., who co-localized glial and neuronal markers on confocal microscopy with the expression of stem cell markers (Nestin and CD133) [7]. PGNT does not harbor cytogenetic alterations, as observed in conventional glial tumors [1,5].

Radiologically, the differential diagnosis includes all tumors that may present with a cystic component and mural nodule, for example, pilocytic

Table 16.1   Histological and Radiological Characteristics of PGNT[a]

Histological Characteristics of PGNT	Radiological Characteristics of PGNT
Moderate cellularity without necrosis.	Well circumscribed, with or without a capsule.
Microvascular proliferation.	Located in any lobe, with a slight predilection for frontotemporal.
Mitotic figures.	Cystic component (75%–90%).
A biphasic tumor cell population in the form of pseudopapillae and solid areas.	Mural nodule (25%).
These solid areas mainly consist of neuronal elements, which may be in the form of an admixture of small, intermediate, or large cells in a fine fibrillary matrix (neuropil).	May be indistinguishable from diffuse glioma. Location described has included cortex, deep white matter, and deep gray structures.
A single or pseudostratified layer of glial cells overlying hyalinized vessels with interpapillary regions of neurocytic or ganglion cells.	May be periventricular (lateral ventricle) (70%). Intraventricular (4%).
Hemosiderin, Rosenthal fibers, peripheral eosinophilic granular bodies, and calcification have also been noted.	Commonly enhance with contrast (40%–50%).
MIB-1 labeling index predominantly low (0.5%–2.5%, mean 1.3%) [6]. A small number of cases have shown high mitotic and MIB-1 indices without clear correlation to prognosis.	Peritumoral edema: • 55% will not have. • 25%–30% will have minimal to mild. • 10%–15% moderate to severe.
They have been designated as WHO Grade I based on their indolent behavior after surgical resection.	Occasional calcification (25%).

[a] PGNT, papillary glioneuronal tumor.

astrocytoma, ganglioglioma, dysembryoplastic neuroepithelial tumor (DNET), and ependymoma.

Histologically, the differential diagnosis includes ganglioglioma and central neurocytoma. The presence of pseudopapillary structures and hyalinized blood vessels are suggestive of PGNT, as described by Komori [2]. However, such hyalinized vessels can be encountered in other glioneuronal tumors, such as ganglioglioma and central neurocytoma. These differential diagnoses also contain monotonous neurocyte-like cells and can be difficult to distinguish, although they both lack pseudopapillary patterns, which are common in PGNT. PGNT has fewer and smaller ganglion cells mixed with astrocytic glial components, unlike a ganglioglioma. PGNT is also less vascular, as was the case in both of our reported cases. Furthermore, PGNT has fewer mature synapses than other glioneuronal tumors [1,5,6,7].

In contrast, in differentiating from other papillary tumors—such as papillary ependymoma, papillary meningioma, choroid plexus papilloma, choroid plexus carcinoma, and metastatic adenocarcinoma—PGNT can be distinguished due to its location and EMA and cytokeratin immunostaining. Ependymomas usually contain a high amount of fibrillarity, which PGNT lacks.

16.7 Treatment and prognosis

Gross total resection is usually possible, to control symptoms and to establish a histological diagnosis. Pimentel et al. studied the value of the extension of surgical resection and concluded that surgical removal was the main prognostic factor [3].

The usual biological behavior of PGNT is not aggressive. There have been at least three recurrences reported despite gross total resection (GTR), in cases of atypical PGNT histology, and adjuvant treatment has been used, including temozolomide and fractionated radiotherapy [5,8]. These include a 13-year-old with a cystic left frontal tumor, treated with GTR and no adjuvant therapy. This patient presented 4 years later with seizures and focal as well as distal intracranial recurrence. She underwent a biopsy, was treated with focal fractionated radiotherapy plus temozolomide, and was reported to be well at 40 months follow-up. The other was a 7-year-old with a left parietal cystic tumor, treated with GTR and no adjuvant therapy. She presented 3 months later with weakness and tumoral hemorrhage, underwent a second subtotal resection with no adjuvant treatment, and was reported to be well 20 months after surgery [8].

Furthermore, at least seven cases have been given a high proliferation index, but this does not necessarily correlate with a bad outcome, as survival has been over 5 years in some of these cases [3,5]. Radiotherapy or chemotherapy is considered if anaplastic [9,10] or recurrent [8].

There has been one report of death 9 months after surgery. This 75-year-old patient presented with headaches, had a right cystic temporoparietal tumor, and underwent subtotal resection due to proximity to the motor strip; there was no adjuvant treatment. She died due to disease progression. The histology was characterized as atypical PGNT with an MIB-1 of 14.7%. Its immunohistochemical staining profile did not stand out from the other three cases reported in the same series, where the patients were well at 34–48 months of follow-up [5].

The role of vascular endothelium hyperplasia, necrosis, high mitotic rate, and high MIB-1 index is still debatable, and as such the role of histological atypia is also unclear. Longer follow-up data will be required to delineate the character of atypia in PGNT. No clear recommendations exist regarding follow-up, yet to acquire these desired long-term survival data, annual review is prudent; atypical cases with a higher proliferative index merit more frequent multidisciplinary discussion and review. In our cases, we stopped anticonvulsants after 1 year, with no complications since, and continue on yearly radiological and clinical review.

16.8 Conclusion

PGNT forms a good example of a newly diagnosed tumor category in evolution. New classifications and reclassifications of broad categories of brain tumors will hopefully lead to a narrower (i) diagnostic, (ii) prognostic, and (iii) therapeutic profile. PGNT is indolent in the vast majority of cases, with a good response to treatment, but the even rarer presence of atypia is a cautionary call for longer follow-up to help further elucidate its biological behavior.

References

1. Louis DN, Ohgaki H, Wiestler OD, et al. The 2007 WHO classification of tumours of the central nervous system. *Acta Neuropathol.* 2007;114:97–109.
2. Komori T, Scheithauer BW, Anthony DC, et al. Papillary glioneuronal tumor: A new variant of mixed neuronal-glial neoplasm. *Am J Surg Pathol.* 1998;22:1171–83.
3. Pimentel J, Barroso C, Miguéns J, Firmo C, Antunes JL. Papillary glioneuronal tumor-prognostic value of the extension of surgical resection. *Clin Neuropathol.* 2009;28(4):287–94.
4. Pages M, Lacroix L, Tauziede-Espariat A, et al. Papillary glioneuronal tumors: Histological and molecular characteristics and diagnostic value of SLC44A1-PRKCA fusion. *Acta Neuropathol Commun.* 2015;3:85.
5. Myung JK, Byeon SJ, Kim B, et al. Papillary glioneuronal tumors: A review of clinicopathologic and molecular genetic studies. *Am J Surg Pathol.* 2011;35:1794–805.
6. Chen L, Piao YS, Xu QZ, Yang XP, Yang H, Lu DH. Papillary glioneuronal tumor: A clinicopathological and immunohistochemical study of two cases. *Neuropathology.* 2006;26(3):243–8.
7. Govindan A, Mahadevan A, Bhat DI, et al. Papillary glioneuronal tumor- evidence of stem cell origin with biphenotypic differentiation. *J Neurooncol.* 2009;95(1):71–80.

8. Javahery RJ, Davidson L, Fangusaro J, Finlay JL, Gonzalez-Gomez I, McComb JG. Aggressive variant of a papillary glioneuronal tumor. Report of 2 cases. *J Neurosurg Pediatr.* 2009;3(1):46–52.

9. Newton HB, Dalton J, Ray-Chaudhury A, Gahbauer R, McGregor J. Aggressive papillary glioneuronal tumor: Case report and literature review. *Clin Neuropathol.* 2008;27(5):317–24.

10. Vaquero J, Coca S. Atypical papillary glioneuronal tumor. *J Neurooncol.* 2007;83(3):319–23.

17

Rosette-Forming Glioneuronal Tumor

Tayfun Hakan

17.1 Definition

Rosette-forming glioneuronal tumor (RGNT) is a Grade I tumor, which is classified among neuronal and mixed-glial tumors category of the central nervous system by the World Health Organization in their recent classification [1]. First described as a distinct variant of mixed glioneuronal tumor in 1998, RGNT is a rare, slow-growing neoplasm of the fourth ventricular region, preferentially affecting young adults and composed of a biphasic cytoarchitecture including two distinct histologic components. The first component has uniform neurocytes forming rosettes somewhat resembling Homer Wright rosettes and/or perivascular pseudorosettes; the other is astrocytic in nature and resembles pilocytic astrocytoma (Figure 17.1) [2]. In intraoperative squash smears, the biphasic character of these tumors may cause misdiagnosis due to sampling error; only one part of the tumor may be represented in the smear preparation.

The tumor is soft, gelatinous, generally well demarcated, and has a pinkish-gray color at surgery. Some tumors may show infiltration.

Immunohistochemical studies show glial fibrillary acidic protein and S-100 positivity in the astrocytic component and neuron-specific enolase positivity in the neurocytic cells (Figure 17.2). RGNT has a low cellularity and Ki-67 labelling index (1% and below) and does not consist of mitosis, atypia, necrosis, or vascular proliferations [3].

There are a few reports that suggest RGNT may be seen as a component of a tumor, such as a dysembryoplastic neuroepithelial tumor, NF-1, or neurocytoma [4].

17.2 Biology

RGNT is regarded as an indolent tumor with benign biological behavior. As a rare and slow-growing tumor affecting mainly young adults, RGNT typically

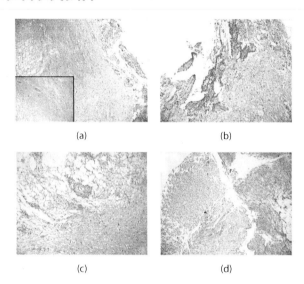

(a)   (b)

(c)   (d)

Figure 17.1   Two components: neurocytic and astrocytic (hematoxylin and eosin [H&E] 40×). (a) The pilocytic astrocytoma–like component on the left side (H&E 200×). (b) Neurocytic pseudorosettes and rosettes around the astrocytic component. (c) Neurocytic pseudorosette: delicate cell processes radiating toward a capillary (H&E 100×). (d) The other glial area resembling oligodendroglioma intermingles with the pilocytic astrocytoma–like area (H&E 100×). There is no mitosis or atypical histopathological findings. (From Hakan, T., and Aker, F.V., *Folia Neuropathol.*, 54, 80–87, 2016.)

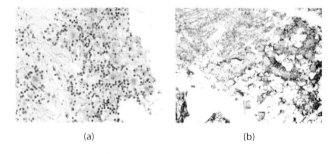

(a)   (b)

Figure 17.2   In an immunohistochemical study, (a) synaptophysin is present at the center of neurocytic rosettes and (b) glial fibrillary acidic protein immunoreactivity is present in the glial component. (From Hakan, T., and Aker, F.V., *Folia Neuropathol.*, 54, 80–87, 2016.)

occurs in the midline, in the brainstem, vermis, pineal region, and/or thalamus. It was previously thought to be specific to the fourth ventricle and/or posterior fossa, but an increasing number of reported cases have confirmed that RGNTs can arise from outside of the posterior fossa: chiasma, suprasellar region, pineal area, septum pellucidum, intraventricular dissemination, lateral ventricle, temporal lobe, and spinal cord [5]. RGNT may show a multifocal intraventricular growth pattern and dissemination throughout the supratentorial region.

Some RGNTs may have a high propensity for significant adherence to the ventricular wall and/or infiltration into surrounding structures. García Cabezas et al. [6] reported a highly aggressive case of RGNT of the fourth ventricle with a histopathological diagnosis of RGNT with atypical microvascular proliferation and focal necrosis; the MIB-1 index was 6%. The lesions were located in the cerebellar hemispheres, cerebellopontine angle, and spinal cord of the patient.

Chiba et al. [4] reported two unusual cases of RGNT with different biological features. One case was marked by early onset in life and a relatively high MIB-1 index, ranging between 3.9% and 8.0%, which was slightly higher, whereas the other case was characterized by relatively fast tumor growth. An MIB-1 index that shows tumor proliferation ranges between 0.35% and 3.07% (mean, 1.58%) for pilocytic astrocytoma or RGNT.

There was some evidence of a tendency to intratumoral bleeding following trauma in two reported cases located in the fourth ventricle. The patients were a young male and an elderly male, in whom intralesional hemorrhage was detected after mild traumatic head injury. The young patient did well, but the elderly one died due to a series of complications.

17.3 Epidemiology

The average age at the time of diagnosis for RGNT is approximately 29 (range: 4–79 years), and this tumor preferentially affects young female adults [3]. Two reports of RGNT show the wide age range—one in a 4-year-old girl [4] and one in a 79-year-old male.

17.4 Pathogenesis

RGNT is thought to originate from progenitor pluripotential cells of the subependymal plate, capable of differentiating along both glial and neurocytic lines. The tumor is linked to the loss of 1p and gain of 1q, as well as gain of the whole chromosomes 7, 9, and 16. Local amplifications in 9q34.2 and 19p13.3 (encompassing the SBNO2 gene) and the presence of the KIAA1549:BRAF gene on chromosome 7 have been identified. Other mutations concern the IDH1 and IDH2, MLL2, CNNM3, PCDHGC4, SCN1A, and FGFR1 genes. MAPK pathway and methylome changes, driven by KIAA1549:BRAF fusion and MLL2 mutation, respectively, could be associated with the development of this rare tumor entity [7,8].

17.5 Clinical features

RGNT often induces nonspecific clinical signs of acute or chronic nature. Most of the symptoms may be due to obstructive hydrocephalus. The development of hydrocephalus can be chronic and may be clinically asymptomatic.

The most common symptoms are headache, ataxia, and vertigo [3]. Some specific symptoms may be seen depending on the localization of the tumor, including balance disturbance, eye pain and blurred vision, diplopia, ptosis, dysarthria, seizure, memory disorders, dizziness, vertigo, nausea, vomiting, clumsy gait, and neck pain and rigidity. Increased intracranial pressure may cause lethargy, somnolence [3], loss of consciousness, and anisocoria in the patients. Some RGNT patients may be asymptomatic and only diagnosed incidentally [4].

17.6 Diagnosis

The clinical and radiological features associated with RGNT are nonspecific.

In MRI and CT examinations, RGNT is relatively circumscribed and heterogeneous and may have some calcifications, satellite lesions, and cystic components with minimal perilesional edema or mass effect [3]. It is usually a solid–cystic tumor but can be entirely solid or cystic. The tumor is usually iso- to hypointense on T1-weighted sequences; hyperintense to isointense on T2-weighted sequences, with moderate and heterogeneous enhancement, respectively, on contrast administration (Figure 17.3). The tumor may have no contrast enhancement or may have an irregular or a ring-shaped contrast enhancement that led to the consideration of malignancy. Diffusion MRI may demonstrate facilitated diffusion in RGNT patients.

Low density on nonenhanced CT is also typical of RGNT [3]. The tumor may show evidence of previous intratumoral hemorrhage on MRI; restricted diffusion may be a sign of possible intratumoral hemorrhage.

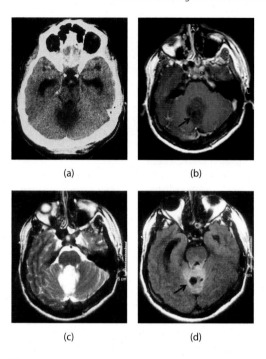

(a)   (b)

(c)   (d)

Figure 17.3 (a) Non-contrast computed tomography scan shows a hypodense lesion at vermis. (b) On T1-weighted magnetic resonance scans, this lesion shows heterogeneous contrast enhancement, and a cystic component is also found (arrow) in the lesion. (c) The lesion seems hyperintense on T2-weighted magnetic resonance imaging (MRI). (d) On Fluid-attenuated inversion recovery (FLAIR)-weighted MRI images, the lesion seems hyperintense, but the cystic component is hypointense (arrow). (From Hakan, T., and Aker, F.V., *Folia Neuropathol.*, 54, 80–87, 2016.)

17.7 Treatment

There is no definite consensus for the management of these tumors. Gross total removal, subtotal removal, and biopsy have all been used for treatment in the literature [3]. As with most other Grade I tumors, RGNT has high cure rates following gross total resection. In general, surgical resection and follow-up observation are recommended for the management of RGNT. However, gross total resection of some RGNTs can be challenging and has significant risks due to their localizations. Major surgical morbidities should be prevented with lesions invading eloquent areas such as the brainstem by performing subtotal resection. Biopsy may be chosen for patients who do not want to risk any postoperative deficit and patients with symptoms that are not severe enough to undertake gross total resection.

In cases where there is acute obstructive hydrocephalus accompanying a tumor and causing deterioration in the consciousness level, a ventriculo-peritoneal (VP) shunt operation or emergent external ventricular drainage is strongly advised for cerebrospinal fluid diversion [3]. Endoscopic third ventriculostomy can be performed during tumor removal or biopsy in cases with ventriculomegaly [4].

Radiotherapy is used in the presence of tumor progression on imaging and/or tumor having progressive symptoms consistent with obstructive hydrocephalus. García Cabezas et al. [6] recommend adjuvant therapy with chemotherapy and radiotherapy for aggressive RGNT. Proton radiotherapy may be used to slow rapid enlargement after partial removal.

17.8 Prognosis

RGNT is a low-grade tumor with no histopathological sign of malignancy [3]. RGNT usually has a good evolution, progression-free survival, and an overall survival of 100% at 2 years.

All but one case of tumors showing progression were found in patients who had undergone subtotal resection or biopsy [4]. Up to the present, only two cases of RGNT that ended with death have been reported in the literature. A 59-year-old female treated with subtotal tumor resection and radiotherapy (5,500 cGy) died after 45 months, and the outcome was attributed to radiation necrosis. A 79-year-old man with RGNT died following minor head trauma. He had lived without clinical deterioration due to tumor progression for 38 months after his initial manifestations.

References

1. Louis DN, Perry A, Reifenberger G, et al. The 2016 World Health Organization classification of tumors of the central nervous system: A summary. *Acta Neuropathol.* 2016;131(6):803–20.
2. Hainfellner JA, Scheithauer BW, Giangaspero F, Rosenblum MK. Rosetteforming glioneuronal tumour of the fourth ventricle. In: Louis DN, Ohgaki H, Wiestler OD, Cavenee WK, eds. *WHO classification of tumours of the central nervous system.* Lyon, France: IARC Press, 2007. p. 115–6.
3. Hakan T, Aker FV. Rosette-forming glioneuronal tumour of the fourth ventricle: Case report and review of the literature. *Folia Neuropathol.* 2016;54(1):80–7.

4. Chiba K, Aihara Y, Eguchi S, Tanaka M, Komori T, Okada Y. Rosette-forming glioneuronal tumor of the fourth ventricle with neurocytoma component. *Childs Nerv Syst.* 2014;30(2):351–6.

5. Anan M, Inoue R, Ishii K, et al. A rosette-forming glioneuronal tumor of the spinal cord: The first case of a rosette-forming glioneuronal tumor originating from the spinal cord. *Hum Pathol.* 2009;40:898–901.

6. García Cabezas S, Serrano Blanch R, Sanchez-Sanchez R, Palacios Eito A. Rosette-forming glioneuronal tumour (RGNT) of the fourth ventricle: A highly aggressive case. *Brain Tumor Pathol.* 2015;32(2):124–30.

7. Gessi M, Moneim YA, Hammes J, et al. FGFR1 mutations in Rosette-forming glioneuronal tumors of the fourth ventricle. *J Neuropathol Exp Neurol.* 2014;73(6):580–4.

8. Bidinotto LT, Scapulatempo-Neto C, Mackay A, et al. Molecular profiling of a rare rosette-forming glioneuronal tumor arising in the spinal cord. *PLoS One.* 2015;10(9):e0137690.

18
Pineal Parenchymal Tumor of Intermediate Differentiation

Prantik Das

18.1 Definition

Tumors arising in the pineal area are heterogeneous in nature and include the pineal parenchymal tumor (PPT), pineoblastoma, pineocytoma, germ cell tumor, and papillary tumor of the pineal region. Among the pineal tumors, pineal parenchymal tumor of intermediate differentiation (PPTID, WHO Grade II to III) is a relatively rare neoplasm with considerable morphological and histological variations. PPTID is a new addition to the WHO classification of central nervous system tumors that may account for up to 10%–20% of all pineal tumors.

18.2 Biology

The pineal region is defined as an area surrounded by the splenium of the corpus callosum and tela choroidea dorsally, the quadrigeminal plate and midbrain tectum ventrally, the posterior aspect of the third ventricle rostrally, and the cerebellar vermis caudally. Pineal region tumors normally originate in cells located in and around the pineal gland. The principle cell of the pineal gland is the pineal parenchymal cell, or *pinocyte*. They are specialized neurons related to retinal rods and cones. The pinocyte is surrounded by a stroma of fibrillary astrocytes, which interact with adjoining blood vessels to form part of the blood–pial barrier.

The pineal gland is richly innervated with sympathetic noradrenergic input from a pathway that originates in the retina and courses through the suprachiasmatic nucleus of the hypothalamus and the superior cervical ganglion. After simulation, the pineal gland transforms the sympathetic input into hormonal output by producing melatonin. Melatonin has a regulatory effect upon hormones such as luteinizing hormone and follicle-stimulating hormone.

Pineocytoma is well differentiated with small, uniform, mature cells and virtually lacks mitoses. It has numerous large pineocytomatous rosettes.

Pineoblastoma is a highly cellular tumor with frequent mitotic figures, irregular nuclei, and a large nucleus-to-cytoplasmic ratio; it forms pattern-less sheets with necrotic areas and few rosettes. PPTID generally shows moderate cellularity, mild to moderate nuclear atypia and low to moderate mitotic activity. Necrosis and endothelial proliferation are absent, in contrast to pineoblastoma.

18.3 Epidemiology

Pineal tumors account for <1% of all primary brain tumors in Europe and North America but are more common in Asian countries.

As the second most common form of pineal tumors, PPT represents 14%–27% of tumors in the pineal gland. The incidence of PPT subtypes varies significantly in the literature, ranging at 14%–60% for pineocytoma, 45% for pineoblastoma, and 10%–20% for PPTID.

There is no particular sex predominance for PPTID, although some authors have reported higher incidence in women. PPTID is more common in middle-aged patients. The median presentation age is around 33 years, and only 13.8% patients belong to the pediatric group. Overall, patients with PPT have significantly worse survival than patients with other pineal tumors (Table 18.1).

18.4 Pathogenesis

Several hypotheses have been postulated for the tumorigenesis of PPTID, including displaced embryonic tissue, malignant transformation of pineal parenchymal cells, or transformation of surrounding astroglia. No specific genetic mutations have been identified for sporadic pineal region tumors. The pathophysiology of pineal region tumors is mostly a result of the anatomic compression of adjacent structures, although local infiltration of neural structures can lead to symptoms in cases of highly invasive tumors. In some cases, neuroendocrine dysfunction is precipitated by specific factors secreted by the tumor. The clinical correlates of this pathophysiology are described in Section 18.5.

18.5 Clinical features

The clinical signs of pineal area tumors may include raised intracranial pressure and/or focal neurological symptoms related to the presence of the pineal tumor itself. The duration of symptoms before diagno-sis is related to tumor growth velocity. Pineal mass often obstructs the

Table 18.1 Summary of Published Studies on Radiotherapy Treatment of PPTID

Authors	Number of Patients	Total Radiotherapy Patients	Craniospinal Radiotherapy Patients	Dose	Stereotactic	Outcome
Schild et al. [1]	4	4	NA	45–64.8 Gy	Not clear	No specific data
Fauchon et al. [2]	26 Classified as Grades II and III, not PPTID	14	12	Neuroaxis Dose average 31 Gy (range 10–38 Gy)	Yes	Five-year survival 74% and 39% for Grades II and III, respectively
Lutterbach et al. [3]	37 Specimens retrospectively analyzed to classify as PPTIDs	NA	Yes NA	Range 20–75 Gy Median 54 Gy	Yes	Median overall survival 165 months
Stobier et al. [4]	1	1		54 Gy/30 f L	No	Time to progression 84 months
Wanatabe et al. [5]	5	2	3	L 54 Gy CSI 36 Gy, WVI 18		Medial overall survival 94.1 months
Ito et al. [6]	6	6	4	22 Gy/10 f L 54.4 Gy/28 f WB + L + WS 50 Gy/25 f EL + L	Yes	Median event-free survival 39 months
Das et al. [7]	5	5	5	54 Gy in 30 fractions, L	No	Median follow-up 21.4 months, no recurrence

Note: CSI, craniospinal irradiation; EL, extended local; L, local; NA, number unknown or unavailable; PPTID, pineal parenchymal tumor of intermediate differentiation; VNCI, vincristine, nimustine, carboplatin, interferon; WB, whole brain; WS, whole spine; WVI, whole ventricular irradiation.

aqueduct of Sylvius, causing hydrocephalus, which leads to increased intracranial pressure. Typical symptoms are headache, nausea, vomiting, cognitive dysfunction, and incontinence. Lesions in the pineal area sometimes cause compression and invasion of the tectal plate and give rise to Parinaud's syndrome, or *dorsal midbrain syndrome*, with a classical triad of vertical gaze palsy, light-near dissociation of the pupils, and convergence retraction nystagmus.

PPTID has the potential to seed through the craniospinal axis and lead to symptoms like cranial neuropathy, radiculopathy, and cauda equina syndrome.

18.6 Diagnosis

Radiologically, peripheral displacement of pineal calcification may help confirm a tumor of pineal parenchymal origin (Figure 18.1). PPTID often demonstrates intermediate to high signal on T2-weighted images, cystic areas, and contrast enhancement. Both pineoblastoma and PPTID may show local invasion, but pineoblastoma is typically seen in children. PPTID is distinguished from pineocytoma by its large size and focal invasion of adjacent structures. Hemorrhage and cysts are common. PPTID should be considered if a locally invasive enhancing PPT is seen in an adult patient. Given PPTID's potential to seed through the craniospinal axis, a preoperative craniospinal MRI should be considered (Figure 18.1).

Cytologically, the tumor cells are generally fragile, so preservation and diagnostic yield should be optimized by ensuring that cerebrospinal fluid (CSF) samples are brought to the laboratory promptly and prepared without delay. Immunocytochemistry is important for further subtyping (Figure 18.1)

The occurrence of necrosis, the mitotic rate, and the immunohistochemical expression of neurofilament protein are used to classify PPTID as Grade II or III pathologically (between Grade I pineocytoma and Grade IV pineoblastoma). A Grade II tumor has <6 mitoses and is strongly immunopositive for neurofilaments, and Grade 3 has >6 mitoses or <6 mitoses but without strong immunostaining for neurofilaments. The MIB-1 level correlates with the WHO grade, and the MIB-1 index is significantly higher in pineblastoma.

Figure 18.2 illustrates the histopathological characteristics of Grade II PPTID. The tumor appears more cellular than in pineocytoma and the cells have more atypical nuclei, but they are not atypical enough to be labelled *pineoblastoma*. The tumor is synaptophysin positive, indicating a tumor of pineal origin.

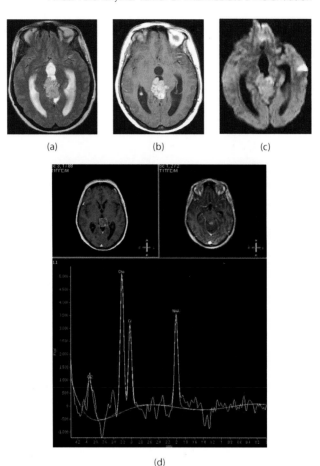

Figure 18.1   T2-weighted MRI. (a) Multiple small cysts within the mass and prob-able invasion of left thalamus, confirmed on T1-weighted scan after intravenous gadolinium. (b) Diffusion-weighted scan (b = 1,000). (c) Small area of diffusion restriction anteriorly within the tumor (confirmed on apparent diffusion coeffi-cient map), close to the left thalamus. (d) Single-voxel MR spectroscopy (TE [echo time] = 144) shows moderate elevation of choline, reduction in N-acetylaspartate (*NAA*), and inverted lactate doublet.

18.7 Treatment

18.7.1 Surgery

Surgery has been the most common treatment utilized, although only one-third of patients may be suitable for a gross total resection. If resection

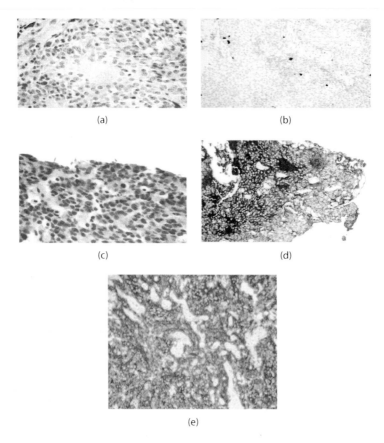

(a)　　　　　　　　　　　(b)

(c)　　　　　　　　　　　(d)

(e)

Figure 18.2　(a) Malignant pineocytes with stippled nuclear chromatin and moderately eosinophilic cytoplasm arranged around a Homer Wright rosette. (b) MIB-1 demonstrates a low proliferative index. (c) Frozen section image: tumor appears more closely packed with denser chromatin. Apoptotic bodies are visible. (d) Synaptophysin immunochemistry shows granular cytoplasmic positivity. (e) Strong neurofilament staining on all cells.

is not possible, biopsy could be considered. Specimens should be fixed in formalin for subsequent histopathological evaluation based on paraffin sections.

The role of surgical debulking in the management of pineal tumors is most clearly defined for pineal tumors that are benign or low grade, when complete surgical resection may be achievable, with excellent long-term recurrence-free survival. Morbidity associated with pineal surgery is high, with reported incidence of 15%–18%.

Endoscopic third ventriculostomy combined (if possible) with transventricular biopsy should be considered as an adjuvant surgical procedure for patients with significant obstructive hydrocephalus. CSF sampling for cytological analysis should be done along with any surgical procedure to alleviate obstruction.

After radical surgery, postoperative MRI should be performed within 48 hours to assess the extent of the residual disease.

18.7.2 Radiotherapy

PPTID is a radiosensitive tumor, with resemblance to pineoblastoma genomically but pineocytoma prognostically. Despite the lack of consensus, a dose of 50.4–54 Gy should be attempted. It is recommended that spinal metastases receive 45–50.4 Gy in 1.8-Gy fractions. Table 18.1 summarizes various published studies using radiotherapy in PPTID [1–7].

18.7.3 Radiosurgery

Stereotactic radiation or radiosurgery (SRS) has been attempted in treating PPT, with a dose range of 12–18 Gy. Patients treated with Gamma Knife with a mean dose of 13.3 Gy or fractionated Cyberknife with 32 Gy have 5-year survival rates of 74% for WHO Grade II PPTID and 39% for Grade III PPTID. There is still ongoing concern about treatment failure in using SRS as a single-modality treatment for more malignant lesions [8].

18.7.4 Chemotherapy

Given the CSF seeding potential of PPTID, there may be a role for systemic chemotherapy in preventing recurrence. One protocol relies on a combined chemotherapy of vincristine, nimustine, carboplatin, and interferon-β to treat PPTID patients after radiotherapy to primary tumor. Another involves cisplatin and vinblastine as systemic treatment of pineal parenchymal cell tumors [9].

Targeted agents, like epidermal growth factor receptor tyrosine kinase inhibitors, could potentially cause more neurotoxic damage compared to conventional chemotherapeutic agents. Radiation may contribute to increased chemotherapeutic neurocognitive toxicity due to blood–brain barrier damage, which increases the penetration of chemotherapy.

18.8 Prognosis

PPTID is associated with a good outcome, with a median progression-free survival of >5 years and median overall survival between 13 and 15 years. Adjuvant radiotherapy has a significant impact on overall survival. Due to the long survival outcome of PPTID patients, late adverse effects are an important consideration when deciding treatment. Use of a wide irradiation field with the addition of systemic therapy invariably increases the long-term side effects of treatment. Late recurrence is also common in PPTID. Thus it is important to keep patients under close follow-up with serial contrast-enhanced MRI in order to detect recurrence. Prolonging follow-up of the patients for at least 10 years is necessary [10].

References

1. Schild SE, Scheithauer BW, Haddock MG, et al. Histologically confirmed pineal tumors and other germ cell tumors of the brain. *Cancer.* 1996;78(12):2564–71.
2. Fauchon F, Jouvet A, Paquis P, et al. Parenchymal pineal tumors: A clinicopathological study of 76 cases. *Int J Radiat Oncol Biol Phys.* 2000;46:959–68.
3. Lutterbach J, Fauchon F, Schild SE, et al. Malignant pineal parenchymal tumors in adult patients: Patterns of care and prognostic factors. *Neurosurgery.* 2002;51(1):44–55 (discussion 55–6).
4. Stoiber EM, Schaible B, Herfarth K, et al. Long term outcome of adolescent and adult patients with pineal parenchymal tumors treated with fractionated radiotherapy between 1982 and 2003—a single institution's experience. *Radiat Oncol.* 2010; 5:122.
5. Watanabe T, Mizowaki T, Arakawa Y, et al. Pineal parenchymal tumor of intermediate differentiation: Treatment outcomes of five cases. *Mol Clin Oncol.* 2014;2(2):197–202.
6. Ito T, Kanno H, Sato K, Oikawa M, Tanaka S. Clinicopathologic study of pineal parenchymal tumours of intermediate differentiation. *World Neuro Surg.* 2014;81(5):783–9.
7. Das P, Mckinstry S, Devadass A, Herron B, Conkey DS. Are we over treating pineal parenchymal tumour with intermediate differentiation? Assessing the role of localised radiation therapy and literature review. *SpringerPlus.* 2016;5:26. doi:10.1186/s40064-015-1502-9.
8. Park JH, Kim JH, Kwon DH, Kim CJ, Khang SK, Cho YH. Upfront stereotactic radiosurgery for pineal parenchymal tumors in adults. *J Kor Neurosurg Soc.* 2015;58(4):334–40.
9. Anan M, Ishii K, Nakamura T, et al. Postoperative adjuvant treatment for pineal parenchymal tumour of intermediate differentiation. *J Clin Neurosci.* 2006;13:965–8.
10. Jouvet A, Saint-Pierre G, Fauchon F, et al. Pineal parenchymal tumors: A correlation of histological features with prognosis in 66 cases. *Brain Pathol.* 2000;10:49–60.

19

Pineocytoma, Pineoblastoma, and Papillary Tumor of the Pineal Region

19.1 Definition

Arising from pinealocytes in the pineal gland, pineal parenchymal tumors (PPT, also called *true pineal cell tumors*) include well-differentiated pineocytoma (PC), pineal parenchymal tumor of intermediate differentiation (PPTID), and poorly differentiated pineoblastoma (PB). In addition to these true pineal cell tumors, papillary tumor of the pineal region (PTPR) is a distinct neoplasm that involves specialized ependymal cells and forms papilla in the pineal region.

PC (WHO Grade I) is a slow-growing, well-differentiated tumor characterized by the abundance of cytoplasmic processes and by the presence of pineocytomatous rosettes. It contains giant cells in addition to areas composed of neoplastic gangliocytes and astrocytes in various combinations. For example, PC with neuronal differentiation shows large atypical ganglion cells and a regular pattern of large mature rosettes (composed of cells with small round nuclei and scanty cytoplasm surrounding delicate fibrillary processes) [1].

PPTID (WHO Grade II to III) sits between PC and PB. PPTID has the potential to metastasize to other regions (e.g., the thoracic and lumbosacral spinal region) (see Chapter 18) [2].

PB (WHO Grade IV) is a fast-growing, poorly differentiated tumor characterized by the scarcity of cytoplasmic processes and by the presence of Homer Wright rosettes. PB consists of giant cells and spreads via the cerebrospinal fluid [3].

PTPR (WHO Grade II to III) is an epithelial-appearing lesion with focal papillary architecture and other densely cellular areas that show ependymal-like differentiation, true rosettes, and tubules consisting of cells with a clear or vacuolated cytoplasm.

19.2 Biology

The pineal gland is a small midline structure (with dimensions of 7 × 6 × 3 mm and a weight of ~175 mg) located behind the third ventricle and between the two hemispheres of the brain. The main cells of the pineal gland are pinealocytes, which secrete melatonin (a hormone involved in the regulation of circadian rhythms). Pinealocytes may be distinguished into two types (I and II). Type I pinealocytes (or *light pinealocytes*) are round or oval-shaped cells of 7–11 μm in diameter and contain the neurotransmitter serotonin, which is later converted to melatonin. Type II pinealocytes (or dark pinealocytes) are round, oval, or elongated cells of 7–11.2 μm in diameter, have many infoldings in the nuclei, and contain melatonin.

Pineal parenchymal tumors mostly originate from pinealocytes, with pineocytoma, PPTID, and pineoblastoma corresponding to the mature, intermediate, and immature forms of the disease, respectively.

Interestingly, pineoblastoma was previously considered part of embryonal tumors (including medulloblastoma, atypical teratoid/rhabdoid tumor, PB, ependymoblastoma, cerebral neuroblastoma, ganglioneuroblastoma, medulloepithelioma, and supratentorial embryonal tumor) within the primitive neuroectodermal tumor category. The 2016 WHO classification of central nervous system tumors groups pineoblastoma with pineocytoma, PPTID, and PTPR as tumors of the pineal region.

PTPR likely develops from papilla-forming ependymal cells, which line the inside of the ventricles of the brain and which contain proteins originating from blood vessels, nerve cells, and muscle cells.

19.3 Epidemiology

Tumors of the pineal region account for <1% of all primary brain tumors and often affect young adults of 20–40 years in age. However, about 3%–8% of childhood brain tumors are found in this area.

Pineocytoma represents 45% of pineal parenchymal tumors and occurs in people aged 1–39 years (mean age of 12.6 years, and median age of 30 years at diagnosis). There is a slight male predilection (52%).

PPTID is responsible for 10% of pineal parenchymal tumors and occurs at all ages, with peak incidence in early adulthood.

Pineoblastoma accounts for 45% of pineal parenchymal tumors and is more commonly identified in children (90%) than in adults (10%), with peak incidence in the first 4 years of life.

PTPR occurs in patients of 5–66 years (mean age of 31.5 years). There is a predominance of female patients.

19.4 Pathogenesis

Pineocytoma is linked to loss of chromosomes 11 and 22 and deletion in the distal 12q region.

Pineoblastoma often contains germline mutations in the retinoblastoma (*RB1*) gene on the long (q) arm of chromosome 13 at Position 14.2 (i.e., 13q14.2) and *DICER1* on the long (q) arm of chromosome 14 at Position 32.13 (i.e., 14q32.13), leading to uncontrolled cell growth and tumorigenesis [4,5].

PTPR shows frequent loss of chromosomes 3, 10 (which contains the phosphatase and tensin homolog), 14, 22, and X and gain of chromosomes 8, 9, and 12.

19.5 Clinical features

Symptoms of pineal parenchymal tumors range from nausea/vomiting, seizures, memory disturbances, to progressive diplopia (double vision), as a result of a growing tumor mass that blocks the cerebrospinal fluid flow and disturbs the eye movement pathways, leading to hydrocephalus.

Patients with pineocytoma tend to show headache (75%), nausea (23%), hydrocephalus (65%), and visual changes (17%) as well as Parinaud syndrome.

Pineoblastoma often causes headache, vomiting, hydrocephalus, blurred/double vision, eye pain, hearing impairment, upward gaze paralysis, and altered sleeping patterns.

The most common symptoms of PTPR are headache and visual disturbances.

19.6 Diagnosis

Pineocytoma is a slow-growing, well-demarcated tumor of <3 cm with occasional cystic change. The tumor appears iso to hyperattenuating on CT, hypo to isointense on T1, and isointense on T2, with solid components vividly enhanced on T1 C+ (Gd). Microscopically, the tumor shows benign cells with uniformly sized nuclei, regular nuclear membrane, and light chromatin, together with typical pineocytomatous pseudorosettes, which are irregular circular/flower-like arrangements of cells with a large meshwork of fibers (neuropil) at the center. The tumor stains positive for synaptophysin, neuron-specific enolase (NSE), chromogranin A, PLAP, Ki-67, and beta tubulin III.

Pineoblastoma is a large, lobulated mass with infrequent calcification. On MRI, the tumor is characterized by hypointensity or hypo to isointensity on T1-weighted images, isointensity or iso to hyperintensity on T2-weighted images, and heterogeneous enhancement in the pineal region. Histologically, the tumor is highly cellular with contains darkly stained, small, round, poorly differentiated cells in patternless sheets or aggregates. The cells show high-grade (anaplastic/undifferentiated) features, including hyperchromatic oval nuclei and scanty cytoplasm, and are partially arranged in Homer Wright or Flexner–Wintersteiner rosettes. Mitosis and areas of necrosis are present. The tumor stains positive for NSE, synaptophysin, retinal S-antigen, chromogranin A, and CD57 but negative for glial fibrillary acidic protein (GFAP), myoglobin, HHF35, S-100, and CD99 [6].

PTPR is a large, well-circumscribed tumor of 2–4 cm, with occasional cystic elements. On MRI, PTPR shows hypointensity in T1-weighted sequence, hyperintensity in T2-weighted sequence and enhancement with contrast. Histologically, PTPR is characterized by epithelial-appearing areas with focal papillary architecture and more densely cellular areas with ependymal-like differentiation. Within the cellular areas, true rosettes and tubules (consisting of cells with a clear or vacuolated cytoplasm) are present. Regular, round to oval nuclei containing stippled chromatin and columnar to cuboidal cytoplasm with a well-defined cytoplasmic membrane are observed. Mitotic activity and necrotic foci may be identified. PTPR Grade II grows relatively slowly, has a slightly abnormal microscopic appearance, and may spread into nearby normal tissue. PTPR Grade III is malignant, has actively reproducing abnormal cells, and often grows into nearby normal brain tissue. PTPR stains positive for cytokeratins (e.g., KL1, AE1/AE3, CAM5.2, CK18) in the papillary structures but only focal GFAP staining. It also expresses vimentin (around tumor cell cytoplasm adjacent to vessel walls), S-100 protein, NSE, microtubule-associated protein 2, neural cell adhesion molecule (N-CAM), transthyretin, and surface epithelial membrane antigen [7].

19.7 Treatment

Surgical removal is recommended for pineocytoma, which is gener-ally a well-circumscribed lesion [8]. Nonetheless, PC may recur many years after seemingly successful treatment with craniospinal radiation therapy [9].

Treatment options for pineoblastoma include surgery (to remove as much of the tumor as possible), radiotherapy (to destroy any cancer cells that may remain near the original location of the tumor), and chemotherapy (to kill any cancer cells that may have traveled to other parts of the body) [8]. In PB patient with partial resection, postoperative radiotherapy, compris-ing prophylactic craniospinal irradiation at a dose of 34.2 Gy followed by a local 25.3-Gy 'boost' to the tumor site for a total dose of 59.5 Gy, has proven effective, leading to the complete regression of the tumor without neurological deficits [3].

PTPR may be treated by surgery alone, surgery in conjunction with radio-therapy, and a combination of surgery, radiotherapy, and chemotherapy.

However, PTPR is prone to recur (58% at five years and 70% at six years) after surgery, radiotherapy, and/or chemotherapy [10].

19.8 Prognosis

Pineocytoma is a benign pineal parenchymal tumor with a relatively good prognosis and a 5-year survival rate of 86%. The tumor does not usually regrow/recur after complete resection, although it may sometimes come back with radiation treatment only. This highlights the value of long-term follow-up of PC patients undergoing only radiotherapy [9].

Pineoblastoma is a malignant tumor with poor prognosis. Five-year survival stands at >50% in children with localized disease at diagnosis undergoing aggressive resection, and much lower in patients with disseminated disease at the time of diagnosis. Compared to children, adults with pineoblastoma often have a relatively poor outcome.

PTPR has 5-year estimates of overall and progression-free survival at 73% and 27%, respectively. Incomplete resection correlates with decreased survival and increased recurrence. PTPR also shows occasional spinal dissemination.

References

1. Clark AJ, Sughrue ME, Ivan ME, et al. Factors influencing overall survival rates for patients with pineocytoma. *J Neurooncol.* 2010;100(2):255–60.

2. Patil M, Karandikar M. Pineal parenchymal tumor of intermediate differentiation. *Indian J Pathol Microbiol.* 2015;58(4):540–2.

3. Ai P, Peng X, Jiang Y, Zhang H, Wang S, Wei Y. Complete regression of adult pineoblastoma following radiotherapy: A case report and review of the literature. *Oncol Lett.* 2015;10(4):2329–32.

4. de Kock L, Sabbaghian N, Druker H, et al. Germ-line and somatic DICER1 mutations in pineoblastoma. *Acta Neuropathol.* 2014;128(4):583–95.

5. Goschzik T, Gessi M, Denkhaus D, Pietsch T. PTEN mutations and activation of the PI3K/Akt/mTOR signaling pathway in papillary tumors of the pineal region. *J Neuropathol Exp Neurol.* 2014;73(8):747–51.

6. PathologyOutlines.com. Pineoblastoma. http://www.pathologyoutlines.com/topic/cnstumorpineoblastoma.html. Accessed November 15, 2016.

7. Fèvre Montange M, Vasiljevic A, Champier J, Jouvet A. Papillary tumor of the pineal region: Histopathological characterization and review of the literature. *Neurochirurgie.* 2015;61(2–3):138–42.

8. PDQ Pediatric Treatment Editorial Board. *Childhood Central Nervous System Embryonal Tumors Treatment (PDQ®): Health Professional Version.* PDQ Cancer Information Summaries. Bethesda, MD: National Cancer Institute (US), 2002.

9. Gomez C, Wu J, Pope W, Vinters H, Desalles A, Selch M. Pineocytoma with diffuse dissemination to the leptomeninges. *Rare Tumors.* 2011;3(4):e53.

10. Fauchon F, Hasselblatt M, Jouvet A, et al. Role of surgery, radiotherapy and chemotherapy in papillary tumors of the pineal region: a multicenter study. *J Neurooncol* 2013;112: 223–31.

20
Medulloblastoma

20.1 Definition

Medulloblastoma is a fast-growing malignant tumor (WHO Grade IV) that is separated by the 2016 WHO classification of central nervous system (CNS) tumors into three categories: (i) *medulloblastoma genetically defined* (including medulloblastoma WNT-activated; medulloblastoma sonic hedgehog (SHH)-activated and TP53 mutant; medulloblastoma SHH-activated and TP53 wild type; medulloblastoma non-WNT/non-SHH Group 3; and medulloblastoma non-WNT/non-SHH Group 4); (ii) *medulloblastoma histologically defined* (including medulloblastoma classic; medulloblastoma desmoplastic/nodular; medulloblastoma with extensive nodularity; medulloblastoma large cell/anaplastic), and (iii) *medulloblastoma NOS* (not otherwise specified) [1].

Within the *medulloblastoma genetically defined* category, the *medulloblastoma WNT-activated* subset accounts for around 10% of medulloblastoma tumors and is essentially a histologically classic medulloblastoma with WNT signaling-gene expression signature, β-catenin nuclear staining, and infrequent metastasis at diagnosis.

The *medulloblastoma SHH-activated* subsets account for around 30% of medulloblastoma tumors and demonstrate a desmoplastic/nodular histology. They harbor mutations in the SHH pathway with or without TP53 alteration (TP53 mutant or TP53 wild-type).

The *medulloblastoma non-WNT/non-SHH Group 3* subset comprises around 25% of medulloblastoma cases and demonstrates either classic or large cell/anaplastic histology, with frequent metastasis at the time of diagnosis and elevated expression of c-MYC.

The *medulloblastoma non-WNT/non-SHH Group 4* subset accounts for 35% of medulloblastoma cases and is either classic or large cell/anaplastic tumor, with metastasis at diagnosis being less common than the Group 3 subset. Molecularly, this subset shows *CDK6* amplification, *MYCN* amplification, and, most characteristically, an isochromosome 17q (i17q) abnormality. The Group 4 subset may be further differentiated into Group 4α (with tandem duplication and elevated expression of synuclein alpha interacting protein [SNCAIP]) and Group 4β (with lower SNCAIP expression but amplification or overexpression of MYCN or CDK6).

The *medulloblastoma histologically defined* category contains tumors that do not have characteristic genetic alterations (e.g., WNT, SHH, etc.) and that are defined histologically as classic medulloblastoma, desmoplastic/nodular medulloblastoma, medulloblastoma with extensive nodularity, and large cell/anaplastic medulloblastoma.

The *medulloblastoma NOS* category consists of tumors that do not show the characteristic histopathological features to be classified within the *medulloblastoma histologically defined* category and for which molecular testing is incomplete or unavailable [1].

20.2 Biology

Possibly arising from a common precursor cell of the subependymal matrix in the CNS, embryonal tumors are a heterogeneous group of malignant brain lesions that consist of hyperchromatic cells (blue cell tumors on standard staining) with little cytoplasm, with a high degree of mitotic activity and a tendency to disseminate in the cerebrospinal fluid (CSF) and subarachnoid space.

Previously, embryonal tumors were grouped under the primitive neuroectodermal tumor (PNET), with medulloblastoma being synonymous to infratentorial PNET. Based on tumor location, divergent histopathologic origins, and patterns of differentiation, embryonal tumors were further divided into medulloblastoma, atypical teratoid/ rhabdoid tumor, pineoblastoma, ependymoblastoma, cerebral neuroblastoma, ganglioneuroblastoma, medulloepithelioma, and supratentorial embryonal tumor.

In the 2016 WHO classification of CNS tumors, embryonal tumors include medulloblastoma, embryonal tumor with multilayered rosettes, medulloepithelioma, CNS neuroblastoma, CNS ganglioneuroblastoma, CNS embryonal tumors NOS, atypical teratoid/ rhabdoid tumor, embryonal tumor with rhabdoid features (see Chapter 21); with pineoblastoma being grouped together with pineocytoma, pineal parenchymal tumor of intermediate differentiation, and papillary tumor of the pineal region under tumors of the pineal region (see Chapter 19) [1].

Medulloblastoma usually forms in the posterior fossa of the cerebellum, with 80% of tumors in children found in the vermis of the cerebellum and 50% of tumors in adults involving the cerebellar hemispheres. However, some medulloblastomas may spread to the bone, bone marrow, lung, and other parts of the body.

The *medulloblastoma WNT-activated* subset likely originates from the embryonal rhombic lip region (brain stem). The medulloblastoma SHH-activated subset is believed to emanate from the external granular layer of the cerebellum, and the medulloblastoma non-WNT/non-SHH Group 4 subset may also come from the upper rhombic lip.

20.3 Epidemiology

Medulloblastoma is the most common embryonal tumor in children of up to 4 years of age and represents an important cause of death in children >1 year in the developed world. Medulloblastoma has a peak incidence rate of 6 cases per million under 9 years of age and falls to <2 cases per million in the 15–19 age group, with 75% of medulloblastoma cases involving patients under 16, and adults being occasionally affected.

Specifically, the *medulloblastoma WNT-activated* subset is primarily observed in older children, adolescents, and adults; the *medulloblastoma SHH-activated* subset is mainly noted in children of <3 years and in older adolescence/adulthood. The *medulloblastoma non-WNT/non-SHH Group 3* subset is present in childhood, including infants, with a male-to-female ratio of 2:1; and the medulloblastoma non-WNT/non-SHH Group 4 subset occurs throughout infancy and childhood and into adulthood, with a male-to-female ratio of 3:1. In addition, medulloblastoma with the desmoplastic/nodular histologic variant is more commonly present in infants than children, although it shows an increasing rate again in adolescents and adults.

20.4 Pathogenesis

The medulloblastoma WNT-activated subset often displays activated WNT signaling, mutation in exon 3 of *CTNNB1*, nuclear expression of β-catenin, loss of chromosome 6 (monosomy 6), and occasional MYC mutation. Other possible genetic alterations include mutations in *TP53*, RNA helicase *DDX3X*, and the chromatin modifiers *SMARCA4* and *MLL2* [2].

The *medulloblastoma SHH-activated and TP53 mutant* subset harbors aberrations in the SHH pathway genes (including *PTCH1*, *PTCH2*, *SMO*, *SUFU*, *GLI2*, and *MYCN*). Other genetic alternations include chromosome 14q and 17p losses, chromosome 9q and 10q deletions, chromosome 3q gain, chromothripsis, p53 amplification, and *TP53* mutation [2].

The *medulloblastoma non-WNT/non-SHH Group 3* subset is associated with amplification of iso17q and *MYC* (of the *MYC* locus on chromosome 8q), as well as TGF-β signaling components and transcription factor *OTX2* (possible target of the TGF-β pathway).

The *medulloblastoma non-WNT/non-SHH Group 4* subset exhibits a higher percentage of iso17q than the Group 3 subset (66% vs. 26%) and utilizes the LIM homeobox transcription factor 1 (LMX1A) as a major regulatory factor. Other genetic alternations include loss of one copy of the X chromosome in female patients and mutations in the histone demethylase KDM6A [2].

Across all medulloblastoma subsets, frequent genetic alterations relate to chromatin regulators (e.g., *MLL2, MLL3*, and *EHMT1; KDM6A, KDM6B, JMJD2C*, and *JMJD2B*; and *SMARCA4, CHD7*, and *ARID1B*) and methylation patterns at and downstream of promoters (e.g., hypomethylation and overexpresssion of the mRNA processing gene *LIN28B* in most Group 3 and Group 4 tumors [2].

Further, a small number of medulloblastoma cases arise in the setting of hereditary cancer predisposition syndromes, including (i) Turcot syndrome (germline mutations in *APC*), (ii) Rubinstein–Taybi syndrome (germline mutations in *CREBBP*), (iii) Gorlin syndrome (germline *PTCH1* and *SUFU* mutations), (iv) Li–Fraumeni syndrome (germline mutations in *TP53*), and (v) Fanconi anemia (mutations in *FANCA-FANCM*) [2].

In contrast to pediatric patients showing focal amplification of *MYC/MYCN*, adult patients often show focal amplification of *CDK6*. Gains of chromosomes 3q, 4, and 19 are common in adult patients, whereas gains of chromosomes 1q, 2, 7, and 17q, as well as loss of 16q, are noted frequently in pediatric patients [2].

20.5 Clinical features

The early symptoms of medulloblastoma are usually related to blockage of CSF and resultant accumulation of CSF in the brain (hydrocephalus) and include abrupt onset of headaches (in the morning on waking), nausea/vomiting, nonspecific lethargy, feeding difficulty, ataxia/truncal unsteadiness, nystagmus, and papilledema.

About 20% of patients with more laterally positioned medulloblastomas do not have hydrocephalus at the time of diagnosis and often present initially with lateralizing cerebellar dysfunction (appendicular ataxia), such as unilateral dysmetria, unsteadiness, and weakness of the sixth and seventh nerves on the same side as the tumor. With the tumor later growing toward the midline and blocking CSF, more classical symptoms associated with hydrocephalus may emerge.

20.6 Diagnosis

Preliminary diagnosis of medulloblastoma is possible through CT and MRI assessment of the entire brain and spine, with and without contrast enhancement (gadolinium). MRI helps better determination of the anatomic relationship between the tumor and the surrounding brain compared to CT scan. Definitive diagnosis of medulloblastoma relies on histological examination of tumor tissue and/or lumbar CSF.

Macroscopically, medulloblastoma is a circumscribed, friable, fleshy, gray–tan tumor of a few centimeters in size with occasional hemorrhage and necrosis, frank invasion of adjacent structures, and infiltration of the meninges and subarachnoid space. The desmoplastic variant may have a firm consistency due to extensive stromal reticulin deposition.

On CT, medulloblastoma appears as hyperdense, noncalcified lesion of the fourth ventricle or cerebellar hemisphere; on MRI, the tumor shows a mass with hypointense appearance on the T1-weighted image, hyperintense appearance on T2-weighted images, and bright enhancement on T1-weighted images after contrast administration.

Histologically, medulloblastoma is defined by the WHO as a "malignant, invasive embryonal tumor of the cerebellum with preferential manifestation in children, predominantly neuronal differentiation, and inherent tendency to metastasize via CSF pathways."

Classic medulloblastoma consists of tightly packed and poorly differentiated small, round, or oval cells with scanty cytoplasm and dense basophilic nuclei. Glial or neuronal differentiation may be present. Perivascular pseudorosettes or Homer Wright rosettes (neuroblastic rosettes of nuclei in a circle around tangled cytoplasmic processes) may be observed in some cases.

Desmoplasitc/nodular medulloblastoma and medulloblastoma with extensive nodularity are characterized by the round, oval, or elongated "pale islands" (nodules) composed of differentiated cells that resemble neurocytes or small mature neurons with variable neuropil formation. In medulloblastoma with extensive nodularity, nodules become extensive or confluent, with minimal proliferative internodular tissue.

Large cell medulloblastoma is characterized by enlarged round cells with giant nuclei and prominent nucleoli, together with abundant apoptotic and mitotic figures, and vesicular chromatin, whereas anaplastic medulloblastoma is characterized by undifferentiated cells with pleomorphic, angular or molded nuclei, together with more pronounced mitotic and apoptotic activity than in classic medulloblastoma. Both large cell and anaplastic medulloblastomas are known for their aggressive behavior and their tendency to metastasize via CSF and outside the CNS [3].

Medulloblastoma staging is based on preoperative MRI of the brain and spine, postoperative MRI of the brain to determine the amount of residual disease, and lumbar CSF analysis.

Risk stratification of medulloblastoma patients of >3 years of age leads to the categorization into those with average risk (having totally resected or near-totally resected [≤1.5 cm2 of residual disease] and no metastatic disease) and those with high risk (having metastatic disease [i.e., neuroradiographic evidence of disseminated disease, positive cytology in lumbar or ventricular CSF obtained more than 10 days after surgery, or extraneural disease] and/or subtotal resection [>1.5 cm^2 of residual disease]). Children with tumors showing diffuse anaplasia are assigned to the high-risk group.

Use of molecular techniques permits improved determination of medulloblastoma, including the delineation of the *medulloblastoma genetically defined* category.

20.7 Treatment

Current treatment for medulloblastoma includes surgery, radiotherapy, and chemotherapy. Medulloblastoma surgery carries a 25% risk of cerebellar mutism syndrome (e.g., delayed onset of speech, suprabulbar palsies, ataxia, hypotonia, and emotional lability). Adjuvant craniospinal radiotherapy may lead to a 2–4 point IQ decrease per year due to radiation-induced brain injury. Chemotherapy consists of cyclophosphamide, etoposide, cisplatin, and vincristine, with or without concomitant high-dose intravenous methotrexate and/or intrathecal methotrexate or mafosfamide, and/or intraventricular methotrexate. Molecularly targeted therapy represents a recent addition to the therapeutic arsenal for medulloblastomas. For example, the SHH *PTCH1* inhibitor vismodegib increases radiographic responses in pediatric patients with recurrent sonic hedgehog (SHH) medulloblastoma subset [4].

20.8 Prognosis

The *medulloblastoma WNT-activated* subset carries a very good prognosis, with a 5-year overall survival rate of >95%, especially in cases with beta-catenin nuclear staining and proven 6q loss and/or *CTNNB1* mutations.

The *medulloblastoma SHH-activated* subset has a 5-year overall survival rate of around 64%. Patients with mutations upstream of the SHH signaling pathway (e.g., *PTCH1*, *PTCH2*, and *SUFU*) have a more favorable prognosis than patients with downstream genomic aberrations (e.g., *GLI2* and *MYCN* amplification). Patients with a large cell/anaplastic histology, metastatic disease, chromosome 14q loss, and mutations in TP53 have a poor prognosis (<50% survival).

The *medulloblastoma non-WNT/non-SHH Group 3* subset has a poor prognosis, with a 5-year overall survival rate of <50% after diagnosis. Children <4 years of age at diagnosis with MYC amplification or *MYC* overexpression have a higher risk of relapse and poorer prognosis than those >4 years of age.

The *medulloblastoma non-WNT/non-SHH Group 4* subset has a better prognosis than the Group 3 subset but not as good as the WNT subset. Prognosis for the Group 4 subset is negatively impacted by the presence of metastatic disease and chromosome 17p loss.

Within the *medulloblastoma histologically defined* category, classic tumor has an average risk, desmoplastic/nodular tumors have a more favorable prognosis, and large-cell/anaplastic tumors have a very poor prognosis.

References

1. Louis DN, Perry A, Reifenberger G, et al. The 2016 World Health Organization classification of tumors of the central nervous system: A summary. *Acta Neuropathol.* 2016;131: 803–20.
2. Kijima N, Kanemura Y. Molecular classification of medulloblastoma. *Neurol Med Chir (Tokyo).* 2016;56(11): 687–97.
3. Borowska A, Jóźwiak J. Medulloblastoma: molecular pathways and histopathological classification. *Arch Med Sci.* 2016;12(3): 659–66.
4. PDQ Pediatric Treatment Editorial Board. *Childhood Central Nervous System Embryonal Tumors Treatment (PDQ®): Health Professional Version.* PDQ Cancer Information Summaries. Bethesda, MD: National Cancer Institute (US), 2002.

21

CNS Primitive Neuroectodermal Tumors and Other Embryonal Tumors

21.1 Definition

Tumors of embryonal origin comprise a diverse group of malignant lesions (all WHO Grade IV) that were previously distinguished into (i) medulloblastoma (also known as *infratentorial primitive neuroectodermal tumor* [infratentorial PNET]), (ii) central nervous system (CNS) PNET (formerly *supratentorial PNET*, consisting of CNS neuroblastoma, CNS ganglioneuroblastoma, medulloepithelioma [MEPL], and ependymoblastoma [EBL]), and (iii) atypical teratoid/rhabdoid tumor (AT/RT).

In the 2016 WHO classification of CNS tumors, embryonal tumors other than medulloblastoma (see Chapter 20) are divided into eight categories: (i) *embryonal tumor with multilayered rosettes* (*ETMR*) chromosome 19 microRNA cluster (C19MC) altered (covering C19MC-amplified lesions, previously known as *embryonal tumors with abundant neuropil and true rosettes* [ETANTR], ETMR, EBL, and some MEPL); (ii) ETMR, NOS (i.e., C19MC-non-amplified lesions with histological features conforming to ETANTR/ETMR); (iii) *MEPL* (*bona fide* MEPL without C19MC amplification); (iv) CNS neuroblastoma; (v) CNS ganglioneuroblastoma; (vi) CNS embryonal tumor, NOS (i.e., tumor previously designated as *CNS PNET*); (vii) AT/RT (defined by alterations of either *INI1* or *BRG1*); and (viii) CNS embryonal tumor with rhabdoid features (referring to a tumor with histological features of AT/RT but without genetic alterations at *INI1* or *BRG1*) [1].

21.2 Biology

Embryonal tumors are biologically heterogeneous lesions that are characterized by the presence of hyperchromatic cells with little cytoplasm, high mitotic activity and apparent cellular transformation; and that are disseminated throughout the nervous system via cerebrospinal fluid pathways [2–4].

The WHO classification of embryonal tumors reflects the presumed location/ origin in the CNS, together with alterations in microRNA cluster C19MC and other chromosomal loci [1].

ETMR C19MC altered category is characterized by a focal amplification at chromosome region 19q13.42 associated with an upregulation of the oncogenic microRNA cluster C19MC (and perhaps *mir-371-373* cluster as well). Within this category, ETMR is a subtype of embryonal tumor with a spectrum of histologic features that led to its description under three different names: ETANTR, EBL, and MEPL. ETMR demonstrates multi-layered (ependymoblastic) rosettes, LIN28A immunoexpression (a highly specific marker), and the 19q13.42 amplification. In addition to display-ing ependymoblastic rosettes (with well-formed central round or slit-like lumina in the absence of an outer membrane) and patterns of neuronal differentiation (including neurocytes, ganglion cells, and neuropil-like background), ETANTR and EBL also show an increased frequency of chromosome 2 gain and a highly specific focal amplification at 19q13.42 involving the C19MC cluster. These put ETANTR and EBL into an identi-cal biological entity within the ETMR C19MC altered category [2–4]. MEPL mimics the embryonic neural tube with external limited mem-brane (the so-called medulloepithelial rosettes). Some MEPL may also display ependymoblastic rosettes. Furthermore, MEPL often shows mul-tiple lines of differentiation including neuronal, glial, and mesenchymal elements. Some MEPL with ETANTR components are found to have 19q13.42 amplification [4].

ETMR with C19MC alteration may occur in both supratentorial (70%) and infratentorial (30%) locations. EBL is found in brain cells lining the fluid-filled spaces in the brain and spinal cord, as well as near the tailbone. MEPL is present in brain cells that line tubelike spaces in the brain and spinal cord. Both EBL and MEPL are fast-growing tumors [2].

CNS embryonal tumor NOS category covers tumors previously designated as CNS PNET, which may display divergent degrees of differentiation along with neuronal, astrocytic, muscular, or melanocytic lines. Tumors with neuronal differentiation are known as CNS neuroblastoma, or CNS gan-glioneuroblastoma (if ganglion cells are present). CNS neuroblastomas are characterized by the presence of Homer Wright (neuroblastic) rosettes, foci of neurocytic and/or ganglion cell maturation, intense synaptophysin expression, and *MYC/MYCN* amplifications in almost 50% of cases. CNS ganglioneuroblastoma differs from CNS neuroblastoma by having chromo-some 17 alterations (loss of 17p or isochromosome 17q [i17q], which are very rare in CNS neuroblastoma) [4].

CNS PNET (often referred to as *noncerebellar PNET*) may arise in the cerebral hemispheres, brainstem, or spinal cord and contains undifferentiated or poorly differentiated neuroepithelial cells with the capacity for differentiation along neuronal, glial, or other lineages. Biologically and clinically, CNS PNET is distinct from the peripheral PNET/extraosseous Ewing sarcoma group of non-CNS tumors.

CNS neuroblastoma develops in the nerve tissue of the cerebrum or the layers of tissue that cover the brain and spinal cord. A lesion containing only neuronal lineage is known as *cerebral neuroblastoma*. CNS neuroblastoma may be large and spread to other parts of the brain or spinal cord.

CNS ganglioneuroblastoma forms in the nerve tissue of the brain and spinal cord. A lesion containing only neuronal lineage is known as *cerebral ganglioneuroblastoma*. CNS ganglioneuroblastoma may be located in one area and fast growing or located in more than one area and slow growing.

AT/RT category contains a distinctive tumor cell type (rhabdoid cell) with vesicular nuclei, prominent nucleoli, and cytoplasmic inclusion-like structures composed of whorls of intermediate filaments [4].

AT/RT may arise in either supratentorial or infratentorial locations, with the cerebral hemispheres being the preferred site above the tentorium and occasional occurrence in the pineal region, septum pellucidum, and hypothalamus. AT/RT demonstrates a variable presence of small cell/embryonal, mesenchymal, and epithelial features, with the embryonal component as well as rhabdoid cells being found often.

21.3 Epidemiology

ETMR with C19MC alteration mainly occurs in children <4 years of age, with a female predilection.

CNS PNET often affects children or young adults, with rare cases involving adults. There is a slight male predominance.

AT/RT accounts for 1%–2% of pediatric brain tumors and is mostly found in children of <3 years, with a male predominance (from 3:2 to 2:1).

21.4 Pathogenesis

MicroRNAs (miRNAs) are small, endogenously expressed nonprotein coding RNAs (19–23 nt in length) that interact with target messenger RNAs and regulate a broad range of cellular and developmental processes. The chromosome 19 microRNA cluster (C19MC) spans ~100 kb at human chromosome 19q13.41 and comprises ~46 tandemly repeated microRNA genes exclusively expressed from the paternally inherited allele in the placenta.

Alteration in the C19MC cluster (with amplicon from 19q13.41–42) is a key marker for ETMR (including ETANTR, EBL, and MEPL), in addition to LIN28A overexpression [3]. MEPL also demonstrates amplification of the *hTERT* gene on 5p15; gain on chromosomal arms 3p, 6p, 14q, 15q, and 20q; and losses on 4q, 5q, 13q, and 18q [4].

CNS PNET harbors loss of the *CDKN2A* locus, with 25% of tumors showing amplification at 19q13.41. Compared to medulloblastoma, supratentorial PNET has a relatively high level of expression of *SOX, NOTCH1, ID1*, and *ASCL-1* transcripts as well as an activation of JAK/STAT3 signaling [4].

CNS neuroblastoma has 13q telomeric deletion, 14q deletion, homozygous deletion of 9p21.3 spanning the *CDKN2A* and *CDKN2B* loci, 19q gain, *RASSF1A* promoter methylation, transcriptional silencing of the *DLC1* gene, expression of *Neuro D* family genes, and FOXR2 activation (CNS NB-FOXR2) [4].

AT/RT contains homozygous deletion, heterozygous deletion, copy-number neutral loss of heterozygosity and mutation affecting each of all nine exons of the putative tumor suppressor gene *INI1(hSNF5/SMARCB1)* located on Chromosome 22q11.2 [4].

21.5 Clinical features

ETMR is associated with increased intracranial pressure, seizures, hemiparesis, cerebellar signs, cranial nerve palsies, and other neurologic deficits.

As CNS PNET lesions are usually large, they often cause an increase in head circumference before closure of the cranial sutures in infants. Other signs include raised intracranial pressure, seizures, and focal neurology.

AT/RT may cause lethargy, vomiting, and failure to thrive. In children >3 years of age, headache and hemiplegia may be observed. Children with posterior fossa tumors may display head tilt and/or cranial nerve palsies.

21.6 Diagnosis

Diagnosis of embryonal tumors such as ETMR, CNS PNET, and AT/RT is helped by suggestive MRI findings, together with histopathologic, immunohistologic, and molecular confirmation.

ETMR is usually a large, demarcated, solid mass featuring patchy or no contrast enhancement, edema, significant mass effect, occasional cystic components, and microcalcifications. Occasionally, cystic components and microcalcification may be observed. On MRI, ETMR shows decreased intensity on T1, increased intensity on T2, and patchy or no contrast enhancement on T1 C+ (Gd). MR spectroscopy reveals choline peak and a high ratio of choline/aspartate (suggestive of hypercellularity of ETMR). Histologically, ETMR is characterized by undifferentiated neuroepithelial cells, abundant well-differentiated neuropil, and characteristic multilayered (ependymoblastic or medulloepithelial) rosettes scattered throughout paucicellular regions of neoplastic neuropil. Molecularly, ETMR C19MC altered is defined by high-level amplification of the microRNA cluster C19MC, gene fusion between *TTYH1* and C19MC, LIN28A overexpression, and frequent trisomy 2. ETMR without C19MC amplification is referred to as ETMR-NOS [7].

CNS MEPL is typically a well-circumscribed, gray–tan mass containing hemorrhage or necrosis, cysts, and calcification. Histologically, the tumor is characterized by an immature tubular, trabecular, or papillary arrangement of neuroepithelial cells that resembles the appearance of the embryonic neural tube, as well as by a range of neoplastic cells that differentiate along neuronal, astrocytic, oligodendroglial, and ependymal lines.

CNS embryonal tumor, NOS (i.e., CNS PNET) is a soft, gray–tan, contrast-enhancing tumor with both solid and cystic components. On MRI, CNS PNET is a contrast-enhancing lesion with areas of calcification as well as low extent of peritumoral vasogenic edema. It appears hypointense to isointense on T1; displays generally high signal solid components, common cystic components, and low signal portions due to calcific components on T2; has markedly heterogeneous enhancement and leptomeningeal seeding on T1 C+ (Gd); and shows restricted diffusion on DWI. MR spectroscopy demonstrates elevated choline, decreased NAA, and elevated taurine (Tau) peak. Histologically, the tumor contains small to medium-sized cells with scanty perinuclear cytoplasm and hyperchromatic nuclei and shows divergent differentiation along neuronal, astrocytic, muscular, or melanocytic lines. Tumors with neuronal differentiation are known as *CNS neuroblastoma*, or *CNS ganglioneuroblastoma* (if ganglion cells are also present). CNS neuroblastoma is characterized by the presence of Homer Wright (neuroblastic) rosettes, foci of neurocytic and/or ganglion cell maturation, intense synaptophysin expression, and *MYC/MYCN* amplification in almost 50% of cases. CNS ganglioneuroblastoma differs from CNS neuroblastoma by having chromosome 17 alterations (loss of 17p or isochromosome 17q [i17q], which are very rare in CNS neuroblastoma). CNS PNET is strongly immunoreactive for INI1 protein and readily distinguishable from INI1-negative AT/RT [8].

AT/RT is typically soft, fleshy, and pink-gray with necrosis or hemorrhage. On MRI, AT/RT may show necrosis, multiple foci of cyst formation, and occasional hemorrhage. It appears isointense to slightly hyperintense to grey matter (especially hemorrhagic areas) on T1; generally hyperintense (hemorrhagic areas can be hypointense) on T2; shows heterogeneous enhancement on T1 C+ (Gd); elevated Cho, and decreased NAA on MR spectroscopy. Histologically, AT/RT displays teratoid (variably mixed epithelioid and spindled components) and rhabdoid features (containing the rhabdoid cells with vesicular nuclei and prominent nucleoli in addition to an eosinophilic inclusion–like cytoplasm composed of intermediate filament whorls). AT/RT is usually reactive for epithelial membrane antigen, vimentin, and smooth muscle actin. Molecularly, AT/RT shows loss of the *SMARCB1* locus, leading to characteristic abrogation of INI1 expression. Some AT/RT do not have *INI1/hSNF5* mutations but contain *BRG1* (the ATPase subunit *SMARCA4* located on 19p13.2) mutations instead. Claudin 6 is another marker that can be used to distinguish AT/RT from other brain tumors [5,9].

21.7 Treatment

Current treatment options for ETMR, CNS PNET, and AT/RT include surgical resection, systemic chemotherapy, craniospinal radiation (when appropriate), and bone marrow or stem cell transplantation [6,10].

21.8 Prognosis

Prognosis for ETMR, CNS PNET, and AT/RT is poor, with ~75% of tumors leading to death within 2 years of diagnosis (average of 9 months for ETMR and 17 months for AT/RT). ETMR with diffuse and widespread immunoexpression of LIN28A has a 5-year overall survival estimate of 0%, whereas other malignant pediatric brain tumors with absence or rare focal positivity of LIN28A have a 5-year survival rate of 68%. The 5-year overall survival rate for CNS PNET patients is 20%–30% and CNS PNET cases with either *MYC* or *MYCN* copy number gain tend to show a worse outcome [6,7].

References

1. Louis DN, Perry A, Reifenberger G, et al. The 2016 World Health Organization classification of tumors of the central nervous system: A summary. *Acta Neuropathol.* 2016; 131(6): 803–20.

2. Ceccom J, Bourdeaut F, Loukh N, et al. Embryonal tumor with multi-layered rosettes: Diagnostic tools update and review of the literature. *Clin Neuropathol.* 2014; 33(1): 15–22.

3. Korshunov A, Ryzhova M, Jones DT, et al. LIN28A immunoreactivity is a potent diagnostic marker of embryonal tumor with multilayered rosettes (ETMR). *Acta Neuropathol.* 2012; 124(6): 875–81.

4. Pfister SM, Korshunov A, Kool M, Hasselblatt M, Eberhart C, Taylor MD. Molecular diagnostics of CNS embryonal tumors. *Acta Neuropathol.* 2010; 120(5): 553–66.

5. Wang J, Liu Z, Fang J, Du J, Cui Y, Xu L, Li G. Atypical teratoid/rhabdoid tumors with multilayered rosettes in the pineal region. *Brain Tumor Pathol.* 2016; 33(4): 261–6.

6. PDQ Pediatric Treatment Editorial Board. *Childhood Central Nervous System Embryonal Tumors Treatment (PDQ®): Health Professional Version.* PDQ Cancer Information Summaries. Bethesda, MD: National Cancer Institute (US), 2002.

7. Radiopedia.org. Embryonal tumors with multilayered rosettes (ETMR). https://radiopaedia.org/articles/embryonal-tumors-with-multilayered-rosettes-etmr; accessed November 25, 2016.

8. Radiopedia.org. Primitive neuroectodermal tumour of the CNS. https://radiopaedia.org/articles/primitive-neuroectodermal-tumour-of-the-cns; accessed November 25, 2016.

9. Radiopedia.org. Atypical teratoid/rhabdoid tumour. https://radiopaedia.org/articles/atypical-teratoidrhabdoid-tumour; accessed November 25, 2016.

10. PDQ Pediatric Treatment Editorial Board. *Childhood Central Nervous System Atypical Teratoid/Rhabdoid Tumor Treatment (PDQ®): Health Professional Version.* PDQ Cancer Information Summaries. Bethesda, MD: National Cancer Institute (US), 2002.

SECTION II
Tumors of the Cranial and Paraspinal Nerves

22
Neurofibroma

Soubhagya Ranjan Tripathy, Samer Hoz Saad Alameri, and
Mohamed Said Elsanafiry

22.1 Definition

Neurofibroma is a benign tumor of the peripheral nerves, resulting
from abnormal proliferation of Schwann cells (cells that cover the nerve
fibers). It is composed of nerve fibers, transformed Schwann cells, blood
vessels, inflammatory white blood cells (mast cells), and connective tissue (fibroblasts, loose material called *extracellular matrix*). Neurofibroma
was previously separated into cutaneous/dermal, subcutaneous, diffuse,
intramuscular, and plexiform (including diffuse plexiform and nodular
plexiform) types. However, in the latest 2016 WHO classification of central
nervous system (CNS) tumors, neurofibroma (WHO Grade I) is divided into
two categories: atypical neurofibroma (which covers all previous nonplexi-
form types) and plexiform neurofibroma (PNF) [1].

22.2 Biology

Schwann cells (or *neurilemma cells*) are a type of glial cells that wrap and
maintain peripheral nerve fibers (axons and dendrites). A typical Schwann
cell looks like a rolled-up sheet of paper, with layers of myelin in between
each coil. Whereas the inner layers are involved in the formation of the myelin
sheath, the outermost layer is nucleated cytoplasm that forms the neurilemma. Although both neurofibroma and schwannoma involve Schwann cells
and contain Antoni A (compact) and Antoni B (loose) areas, they differ in
several aspects.

Firstly, neurofibroma is usually a non-encapsulated, intraneural mass that
engulfs the nerve of its origin (i.e., centripetal) and makes a nerve swell
or balloon out; schwannoma is an encapsulated, extraneural mass that
displaces axons (centrifugal) and sits on top of the nerve (Figure 22.1).
Secondly, neurofibroma tends to have more Antoni B fibers than schwannoma does. Thirdly, neurofibroma is associated with neurofibromatosis type
1 (NF1), whereas schwannoma is linked to neurofibromatosis type 2 (NF2).

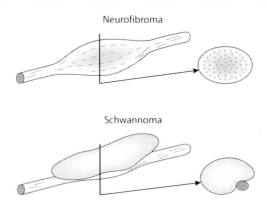

Figure 22.1   Schematic presentation of neurofibromatosis and schwannoma.

Finally, as removal of neurofibroma always means cutting the nerve, surgery for neurofibroma is likely to be more painful than that for schwannoma [2].

22.3 Epidemiology

Neurofibroma affects people of various ages with no sex predilection. Whereas atypical neurofibroma is found in people aged 20–30 years of age, PNF occurs mainly in children and young adults [3,4]. Atypical neurofibroma may develop sporadically as well as in association with NF1, whereas PNF is often congenital and linked to NF1 [5].

22.4 Pathogenesis

Affecting approximately 1 in 2,600–3,000 live births, NF1 is an autosomal dominant genetic disorder due to mutations in the *NF1* tumor suppressor gene located on the long arm of chromosome 17 (17q11.2). The *NF1* gene encodes for neurofibromin, which is a negative regulator of the Ras oncogene. Loss of neurofibromin (as in NF1) results in elevation of growth-promoting signals. The spontaneous mutation rate in the *NF1* gene is high, with 30%–50% of cases representing new somatic mutations. Penetrance is complete, so that germline inactivation of *NF1* is constitutively expressed in all cells [5–8].

Atypical neurofibroma may arise sporadically (nonsyndromic, in the absence of NF1, especially solitary tumor) as well as in association with NF1 (in the case of multiple tumors). As a WHO Grade I neoplasm, atypical neurofibroma is a focal-enhancing mass that abuts but doesn't invade the skull.

PNF is an infiltrative intra- and extraneural neoplasm that occurs exclusively in NF1. Biallelic inactivation of the *NF1* gene has been identified in sporadic PNF. The tumor commonly occurs in the scalp, orbit, pterygopalatine fossa, parotid gland (peripheral branches of CN VII), and ophthalmic branches and rarely in cranial nerves IV [9], V, IX, and X. Extracranial PNF is commonly multicompartmental and not limited by fascial boundaries. Orbital PNF may enlarge the superior orbital fissure (SOF) and extend into the cavernous sinus and as far as Meckel's cave [10].

22.5 Clinical features

Neurofibroma is an intraneural mass that is likely to be more painful than schwannoma. Percussion over a neurofibroma usually produces a dramatic Tinel sign.

Atypical neurofibromas often present as solitary or multiple tumors that are slow growing, small, soft, and painless nodules protruding from the skin. PNF is usually a deeper, larger tumor that causes tortuous enlargement of the peripheral nerves. About 5% of PNF degenerate into malignant peripheral nerve sheath tumors, and >15% of NF1 patients will develop pilocytic astrocytoma. PNF most commonly manifests in children with NF1 as optic pathway glioma (OPG).

As OPG accounts for two-thirds of CNS tumors in patients with NF1, a child diagnosed with OPG should prompt an evaluation for NF1. Up to 15%–20% of children with NF1 harbor an OPG, but only about one-third will develop symptoms. Patients with OPG may display a combination of loss of visual acuity, proptosis, precocious puberty, or visual field deficits. OPG in NF1 generally follow an indolent course; if symptom free beyond puberty, most will remain stable into adulthood or occasionally regress. Extraoptic gliomas can occur. High-grade gliomas are rare.

22.6 Diagnosis

The clinical diagnosis of neurofibroma is based on characteristic cutaneous, ophthalmologic, musculoskeletal, and neurological manifestations. The diagnostic criteria for NF1 is illustrated in Chart 22.1.

As a poorly delineated, wormlike, soft tissue mass that may diffusely infiltrate the scalp, orbit, or parotid gland, neurofibroma shows infiltration and enlargement of soft tissues, typically scalp and periorbital, and appears isodense to muscle on CT. Calcification and hemorrhage are exceedingly

Two or more of the following
• ≥6 café au lait spots, each ≥5 mm in greatest diameter in prepubertal individuals, or ≥15 mm in greatest diameter in postpubertal patients
• ≥2 neurofibromas of any type, or one plexiform neurofibroma (neurofibromas are usually not evident until age 10–15 years). Maybe painful
• Freckling (hyperpignientation) in the axillary or intertriginous (inguinal) areas
• Optic glioma
• ≥2 Lisch nodules: pigmented iris hamartomas that appear as translucent yellow/brown elevations that tend to become more numerous with age
• Distinctive osseous abnormality, such as sphenoid dysplasia or thinning of long bone cortex with or without pseudarthrosis (e.g., of tibia or radius)
• A first-degree relative (parent, sibling, or offspring) with NFI by above criteria

Chart 22.1   Diagnostic criteria of neurofibromatosis type 1.

uncommon. Contrast CT reveals heterogeneous enhancement. Bone window may show expansion of the SOF and pterygopalatine fossa. MRI demonstrates T1 isointensity, T2 hyperintensity with strong heterogeneous enhancement. A target sign of hypointensity, within an enhancing tumor fascicle is seen in some PNF (not pathognomonic though).

Histologically, neurofibroma demonstrates Schwann cells with wirelike collagen fibrils (wavy serpentine nuclei, pointed ends), stromal mucosubstances, mast cells, Wagner–Meissner corpuscles, Pacinian corpuscles, axons, and fibroblasts. Perineural cells are found in plexiform types, with rare mitotic figures, occasional infiltration, and rare skeletal differentiation (neuromuscular hamartoma). Verocay bodies, nuclear palisading, and hyalinized thickening of vessel walls are absent. The tumor stains positive for S-100, CD34 (focal), and Factor XIIIa (focal); but negative for epithelial membrane antigen (except in PNF).

22.7 Treatment

Treatment options for neurofibroma are surgery, radiotherapy, and/or chemotherapy.

22.7.1 Surgery

Atypical neurofibroma may be left alone unless it is growing or causing pain, as it is impossible to completely remove the tumor. PNF can be treated surgically depending on its location and size and whether it is causing problems. PNF of large nerves can sometimes be treated surgically by teasing apart the tumor and trying to remove parts of it without removing the entire nerve. Sometimes large tumors can also be shrunk using a technique called *embolization* to cut off the blood supply to a tumor.

22.7.2 Radiotherapy

External-beam radiotherapy in NF1 should be used cautiously, given the risk for secondary malignancy, neurological dysfunction, cognitive decline, and vascular stenosis.

22.7.3 Chemotherapy

Chemotherapy remains one of the important components of tumor management, with regimens most often including carboplatin with vincristine. Targeting a signaling pathway may help decrease the stimulation/activity of the pathways that are involved in neurofibroma. Intervention should be reserved for those with progressive symptoms or radiographic progression.

22.8 Prognosis

Neurofibroma is a benign nerve sheath tumor. However, there is a risk of malignant transformation in PNF, leading to malignant peripheral nerve sheath tumor. Five-year survival approaches 90% following chemotherapy alone and can substantially delay or defer radiation in children with OPG.

Children with should be screened for OPG with yearly comprehensive ophthalmologic evaluation until puberty, after which screening should be performed every 2 years. In patients with known lesions, radiographic surveillance with gadolinium-enhanced MRI of the brain (with orbits) is recommended but should be combined with ophthalmologic evaluation, because the functional examination, more than radiographic findings, drives management decisions. Pubertal development and growth should be monitored at least annually for signs of precocious puberty.

References

1. Louis DN, Perry A, Reifenberger G, et al. The 2016 World Health Organization classification of tumors of the central nervous system: A summary. *Acta Neuropathol.* 2016;131:803–20.
2. Gutmann DH, Aylsworth A, Carey JC, et al. The diagnostic evaluation and multidisciplinary management of neurofibromatosis 1 and neurofibromatosis 2. *JAMA.* 1997;278(1):51–7.

3. Lammert M, Friedman JM, Kluwe L, et al. Prevalence of neurofibromatosis 1 in German children at elementary school enrollment. *Arch Dermatol*. 2005;141(1):71–4.

4. Walker L, Thompson D, Easton D, et al. A prospective study of neurofibromatosis type 1 cancer incidence in the UK. *Br J Cancer*. 2006;95(2):233–8.

5. Wallace MR, Marchuk DA, Andersen LB, et al. Type 1 neurofibromatosis gene: Identification of a large transcript disrupted in three NF1 patients. *Science*. 1990;249(4965):181–6.

6. Shen MH, Harper PS, Upadhyaya M. Molecular genetics of neurofibromatosis type 1 (NF1). *J Med Genet*. 1996;33(1):2–17.

7. Zhu Y, Ghosh P, Charnay P, et al. Neurofibromatosis in NF1: Schwann cell origin and role of tumor environment. *Science*. 2002;296(5569):920–2.

8. Harrisingh MC, Lloyd AC. Ras/Raf/ERK signaling and NF1. *Cell Cycle*. 2004;3(10):1255–8.

9. Tripathy SR, Mishra SS, Deo RC, et al. Trochlear nerve neurofibroma in a clinically NF1 negative patient: A case report & review of literature. *World Neurosurg*. 2016;89:732.e13–18.

10. Listernick R, Charrow J, Greenwald M, et al. Natural history of optic pathway tumors in children with neurofibromatosis type 1: A longitudinal study. *J Pediatr*. 1994;125(1):63–6.

23

Perineurioma and Malignant Peripheral Nerve Sheath Tumor

23.1 Definition

Arising from perineurial cells within the peripheral nerve sheath, perineurioma (WHO Grade I) is a benign neoplasm with advanced perineurial differentiation, presenting as intraneural perineurioma (a rare, swelling lesion of a major nerve in young adults), soft tissue perineurioma (a common, painless subcutaneous mass on the limbs or trunk of middle-aged adults), and sclerosing perineurioma (a solitary, small nodule in the finger or palm in young adults). It often occurs in the background of neurofibromatosis type 2 (NF2) [1].

Malignant peripheral nerve sheath tumor (MPNST, WHO Grade IV) may arise from the peripheral nerves, from a preexisting peripheral nerve sheath tumor (neurofibroma), or in the setting of neurofibromatosis type 1 (NF1) syndrome (20%–40%), and often demonstrates Schwann cell differentiation on histologic examination [2].

MPNST with heterologous rhabdomyoblastic (skeletal muscle) differentiation is known as *malignant triton tumor* (MTT, a name reflecting the ability of triton salamander to regenerate limbs and to grow both neural and muscle components from a transplanted sciatic nerve) [2].

23.2 Biology

The nervous system of humans comprises two parts: the central nervous system (CNS) and the peripheral nervous system (PNS). The CNS consists of the brain and spinal cord, the PNS encompasses the cranial nerves, spinal nerves (their roots and branches), peripheral nerves, and neuromuscular junctions. The nerves in the PNS connect the CNS to sensory organs (e.g., the eye and ear), muscles, blood vessels, and glands, as well as other organs of the body. Peripheral nerves are covered by an external sheath, which consists of concentric layers of thin perineurial cells.

As a rare nerve sheath tumor with well-differentiated perineurial cells, perineurioma may occur in a variety of anatomic sites. Intraneural perineurioma typically involves spinal nerve roots, trunks, or branches (median, tibial, peroneal, sciatic), solitary (rarely multiple adjacent nerves) and rarely cranial nerves. Characterized by localized, solitary expansion of peripheral nerves, involving one or more nerve fascicles, and by complex perineurial cell proliferation extending into the endoneurium and concentrically surrounding individual nerve fibers and endoneurial capillaries, intraneural perineurioma produces specific pseudo-onion bulbs on cross sections of nerve fascicles. Soft tissue perineurioma is usually well circumscribed with a capsule and contains slender cells arranged in loose fascicles or whorls.

MPNST (formerly *malignant neurilemmoma*, *malignant neurofibroma*, *malignant schwannoma*, *neurofibrosarcoma*, and *neurosarcoma*) occurs in the protective lining of the nerves that extend from the spinal cord into the body, with common presence in the deep tissue of the extremities (45%–59%), trunk (17%–34%), and head and neck (19%–24%).

23.3 Epidemiology

Intraneural perineurioma is a rare tumor with <100 cases described to date. It commonly affects adolescence to early adulthood and shows no gender predilection.

MPNST is a rare disease, with an incidence of 1 in 100,000, accounting for 2% of all soft tissue sarcomas. MPNST may arise sporadically (mainly in the fifth decade, with mean age between 30 and 60 years) or associate with NF1 (the autosomal dominant condition; at an average age of 30 years and mean age of 20–40 years).

MTT typically affects people of 30 years of age. Constituting around 5% of MPNST, MTT is larger (9 cm in size) and portends poorer outcomes than conventional MPNST (6 cm in size).

23.4 Pathogenesis

Perineurioma is linked to NF2, an autosomal dominant disorder resulting from mutations in the *NF2* gene located on chromosome 22 (22q12.2), spanning 93 kb with 7 exons. The *NF2* gene encodes at least 10 protein isoforms including tumor suppressor merlin (moesin-, ezrin-, radixin-like protein; also referred to as *schwannomin*) via a combination of alternative splicing [3].

Alterations in the *NF2* gene are also found in other tumors such as schwannomas, meningiomas, astrocytomas, and ependymomas [4]. Further, the *NF2* gene demonstrates a similarly high rate of *de novo* mutation. Moreover, chromosome 10 aberrations, t(2;10)(p23;q24), and monosomy 10 are noted in sclerosing perineuriomas, whereas chromosome 22 abnormalities (monosomy of chromosome 22) are present in other perineurioma types.

About 50% of MPNST are related to NF1, an autosomal dominant disorder caused by mutations (nonsense, missense, or frameshift) or total loss in the 283 kb fragment containing 60 exons (commonly known as *NF1*) on the long arm of chromosome 17 (17q11.2), leading to alteration in the *NF1*-coded neurofibromin (a 220 kDa protein), RAS hyperactivity, activation of multiple downstream survival and proliferative pathways, and formation of multiple benign neurofibromas, Lisch nodules, and café au lait spots. Of these, neurofibromas can be separated into dermal neurofibromas (neurofibromas arising in the skin, with hormonal responsiveness and virtually no malignant potential) and plexiform neurofibromas (neurofibromas that occur in large, deeply situated nerves or nerve plexuses, without hormonal responsiveness, but with a probability of 8%–13% of undergoing malignant transformation into MPNST [5].

MPNST may also arise sporadically or following radiation therapy. In addition to NF1, other abnormalities implicated in MPNST tumorigenesis include changes in the TP53 tumor suppressor gene (17p13.1) and p16, gains of chromosome arms 7p, 8q, and 17q, and losses of chromosome arms 9p, 11q, 13q, and 17p. Interestingly, whereas *TP53* mutations are often seen in sporadic MPNST, EGFR overexpression and Raf and PI3K/AKT pathway activation are commonly present in NF1-associated MPNST [2].

23.5 Clinical features

Depending on the nerve involved, intraneural perineurioma may manifest as progressive muscle weakness, localized muscle atrophy, nonpalpable mass, pain in affected regions, and sensory disturbance [6].

Commonly occurring in conjunction with large peripheral nerves (e.g., the sciatic nerve, the brachial plexus, and the sacral plexus), MPNST may show an enlarging palpable nodular mass (often in the setting of NF1) and induce various symptoms, ranging from painless swelling in the extremities (arms or legs, also known as *peripheral edema*), soreness and pain (burning or pins and needles), discomfort (numbness), dizziness, and loss of balance to neurological signs [2].

23.6 Diagnosis

Intraneural perineurioma often causes segmental, uniform expansion of the nerve. Microscopically, intraneural perineurioma shows pseudo-onion bulbs surrounding nerve fibers with bundles of spindle-shaped perineurial cells (containing ovoid to elongated nuclei and pale cytoplasm) in longitudinal sections, fine collagenous stroma, irregular borders with the adjacent lamina propria, and entrapped colonic crypts. Perineurioma stains positive for epithelial membrane antigen, Claudin-1, and CD34 (33%) and negative for S-100, glial fibrillary acidic protein (GFAP), neurofilament protein, smooth muscle actin (SMA), desmin, caldesmon, KIT, and pan-keratin. Cytogenetically, perineurioma harbors monosomy 22, and deletion of 22q11-13.1 [7]. Differential diagnosis for intraneural perineurioma includes localized reactive Schwann cell proliferations, while that for soft tissue perineurioma includes low-grade fibromyxoid sarcoma. Notably, low-grade fibromyxoid sarcoma displays prominent stromal collagen deposition and an abrupt transition into myxoid nodules in a curvilinear vascular pattern. However, low-grade fibromyxoid sarcoma is positive for MUC4 expression and a FUS rearrangement, while perineurioma is negative for both.

MPNST is an ovoid or fusiform soft tissue mass along a nerve. On ultrasound, MPNST shows a large, usually elongated, hypoechoic mass; on CT, MPNST has low attenuation (due possibly to fat entrapment, high lipid content of Schwann cell myelin, and cystic areas due to hemorrhage or necrosis); on MRI, MPNST is isointense to muscle on T1 and typically hyperintense on T2 with fascicular appearance; on positron emission tomography (PET)/CT, MPNST appears as fluorodeoxyglucose (FDG)-avid mass [8]. Histologically, MPNST displays infiltration of spindle, rounded, or fusiform cells arranged in intersecting fascicles (which alternate with myxoid regions to form a marbled pattern). In the absence of a history of NF1 or gross or microscopic evidence of nerve sheath tumor or neurofibroma, electron microscopic identification of ultrastructural features of Schwann cells represents the most reliable method of diagnosis. A combination of S-100, Leu-7, and myelin basic protein stains are helpful to confirm the diagnosis of MPNST and to exclude other spindle cell lesions [2].

The MPNST stages (I, II, III, IV) are determined according to the American Joint Committee on Cancer Staging System for Soft Tissue Sarcomas (the TNM system). Stage I is low-grade small soft tissue sarcoma without evidence of metastasis; Stage II is a small (<5 cm) high-grade tumor or large (>5 cm) but superficial high-grade tumor without evidence of metastasis. Stage III is a high-grade large (>5 cm), deep tumor, and Stage IV is a tumor with evidence of metastasis.

MTT is an MPNST occurring in association with rhabdomyosarcoma (or rhabdomyoblastic differentiation, which is characterized by both neural and skeletal muscle differentiation). MTT is differentiated from high-grade MPNST by the presence of cells with skeletal muscle differentiation inside the tumor. Immunohistochemical staining for skeletal muscle markers such as desmin and myogenin provides further confirmation of the identity.

Other examples of MPNST with differentiation include glandular malignant schwannoma, epithelioid malignant schwannoma, and superficial epithelioid MPNST.

Differential diagnoses of MPNST in peripheral nerve and soft tissue include synovial sarcoma, fibrosarcoma, rhabdomyosarcoma, leiomyosarcoma, dedifferentiated liposarcoma and clear cell sarcoma, undifferentiated pleomorphic sarcoma, and angiosarcoma. While MPNST shows partial or complete loss of S100 expression, other benign Schwann cell tumors do not. On the other hand, demonstration of *SS18-SSX1* or *SS18-SSX2* gene fusions, resulting from a characteristic X;18 translocation, may be required for a definitive diagnosis of intraneural synovial sarcoma. Molecular identification of *EGFR* amplification, and *NF1* or *CDKN2A* (p16), offers additional evidence for MPNST.

23.7 Treatment

The first-line treatment for intraneural perineurioma and MPNST is surgical resection with wide margins. In some asymptomatic patients with intraneural perineurioma, treatment may not be necessary. In case of MPNST, preoperative external beam radiation therapy may be administered before surgical resection. If complete removal is impossible, excision combined with high-dose radiation therapy may be employed. Chemotherapy (e.g., doxorubicin, ifosfamide, carboplatin, and etoposide) may be utilized in high-grade, metastatic MPNST and can be administered in preoperative and postoperative settings [5,9,10].

23.8 Prognosis

Intraneural perineuriomas that are small and easily resectable generally have an excellent prognosis. However, for difficult-to-operate tumors, diagnostic biopsy followed by neurolysis instead of resection (or resection with neural grafting or end-to-end anastomosis) may be considered to preserve nerve function. This may likely increase the prognostic uncertainty.

Prognosis is generally poor for patients with MPNST, especially those with large tumors, NF1-association, and the presence of metastasis. The recurrence rate is as high as 40% and about two-thirds of cases metastasize (to the lungs and bone). Five-year survival rates range from 26% to 60%, and 10-year survival is around 45%.

MTT has a worse prognosis than conventional MPNST, with a 5-year survival rate of 11%–15% versus 26%–60%. Thoracoabdominal location of MTT is associated with increased incidence of local recurrence.

References

1. Wang LM, Zhong YF, Zheng DF, et al. Intraneural perineurioma affecting multiple nerves: A case report and literature review. *Int J Clin Exp Pathol*. 2014;7(6):3347–54.
2. Farid M, Demicco EG, Garcia R, et al. Malignant peripheral nerve sheath tumors. *Oncologist*. 2014;19(2):193–201.
3. Schroeder RD, Angelo LS, Kurzrock R. NF2/merlin in hereditary neurofibromatosis 2 versus cancer: Biologic mechanisms and clinical associations. *Oncotarget*. 2014;5(1):67–77.
4. Carroll SL. Molecular mechanisms promoting the pathogenesis of Schwann cell neoplasms. *Acta Neuropathol*. 2012;123(3):321–48.
5. Kamran SC, Shinagare AB, Howard SA, Hornick JL, Ramaiya NH. A-Z of malignant peripheral nerve sheath tumors. *Cancer Imaging*. 2012; 12:475–83.
6. Huang Y, Li H, Xiong Z, Chen R. Intraneural malignant perineurioma: A case report and review of literature. *Int J Clin Exp Pathol*. 2014;7(7): 4503.
7. Basicmedicalkey.com. Intraneural perineurioma. http://basicmedicalkey. com/intraneural-perineurioma/. Accessed December 1, 2016.
8. Rodriguez FJ, Folpe AL, Giannini C, Perry A. Pathology of peripheral nerve sheath tumors: Diagnostic overview and update on selected diagnostic problems. *Acta Neuropathol*. 2012;123(3):295–319.
9. Kolberg M, Høland M, Agesen TH, et al. Survival meta-analyses for >1800 malignant peripheral nerve sheath tumor patients with and without neurofibromatosis type 1. *Neuro Oncol*. 2013;15(2):135–47.
10. Karajannis MA, Ferner RE. Neurofibromatosis-related tumors: emerging biology and therapies. *Curr Opin Pediatr*. 2015;27(1):26–33.

24
Schwannoma

Austin Huy Nguyen, Adam M. Vaudreuil,
Victoria M. Lim, and Stephen W. Coons

24.1 Definition

Schwannomas (also known as *neurilemmomas*) are benign neoplasms comprised of Schwann cells within the nerve sheath. These tumors are typically encapsulated and form discrete masses that sometimes encircle but do not invade the adjoining nerve. Neurofibromatosis type 2 (NF2) and schwannomatosis are conditions in which multiple schwannomas may be present.

24.2 Biology

Schwann cells are the principle cells of the peripheral nervous system (PNS). Derived from the neural crest, the neural crest cells migrate to their eventual target in the PNS during embryologic development. The neural crest cells then differentiate into Schwann cells with the help of various transcription factors, growth factors, proteins, and axonal signaling [1]. The resulting Schwann cells become responsible for the myelination of peripheral axons, with each Schwann cell myelinating a single axon. This process helps provide a protective barrier for the axon and increase salutatory conduction of the neuron. Due to their neural crest origin, schwannomas classically are strongly positive for S-100 protein on immunohistochemical staining.

24.3 Epidemiology

Schwannomas are generally found as sporadic, solitary lesions. Multiple schwannomas may occur in the setting of NF2 or schwannomatosis. There is no gender predisposition, with most reported cases occurring in middle adulthood. These tumors comprise a large percentage of all cerebellopontine angle tumors. However, most schwannomas appear outside the central nervous system (CNS), particularly in the skin and subcutaneous tissues of the head and neck region. The flexural regions of the extremities are another common site. There is a propensity for schwannomas to affect sensory nerves. Those that occur intracranially have a predilection for cranial nerve VIII, especially when associated with NF2. In fact, the presence of bilateral cranial nerve VII schwannomas is diagnostic of NF2. These tumors affect the transition zone between the CNS and PNS of the vestibular portion of the nerve but almost never affect the cochlear portion. Meanwhile, peripheral schwannomas generally affect small cutaneous sensory nerves. Overall incidence of peripheral schwannomas is estimated at 0.6 per 100,000 people annually.

24.4 Pathogenesis

The etiology of schwannomas is unknown, with some theories as to causative factors, including trauma, chronic irritation, and radiation exposure. Vestibular or cranial nerve VIII schwannomas, in particular, are possibly associated with prior radiation exposure [2]. The molecular pathogenesis of schwannomas is largely related to that of NF2. Loss of function or inactivation of the tumor suppressor gene NF2 on chromosome 22q12.2 leads to a decrease in production of the protein merlin. This loss of merlin is a hallmark feature of schwannomas, regardless of NF2 gene status, and is implicated in their tumorigenesis. In schwannomatosis, tumors exhibit NF2 inactivation whereas nontumoral tissue does not, suggesting an alternate genetic cause for schwannomatosis.

24.5 Clinical features

Schwannomas in NF2 characteristically involve the eighth cranial nerve in the cerebellopontine angle (Figure 24.1), classically with bilateral tumors. These present with associated cranial nerve symptoms, including hearing loss, tinnitus, and vertigo. Intraspinal tumors most frequently affect sensory nerve roots, over motor and autonomic nerves. CNS loci are rare. The majority of schwannomas involve peripheral nerves, particularly in the skin and subcutaneous tissue. Anatomically, the head and neck region or flexor surfaces of the extremities are most often affected. Peripheral schwannomas will often present as asymptomatic lesions found incidentally. Paraspinal nerve tumors may cause radicular pain and compression of nerve roots or the spinal cord. Due to the predilection for sensory nerve roots, motor symptoms are uncommon.

24.6 Diagnosis

The overwhelming majority of schwannomas are encapsulated masses measuring less than 10 cm. They are typically of light tan color with intermittent bright yellow cysts; hemorrhage and necrosis may be present.

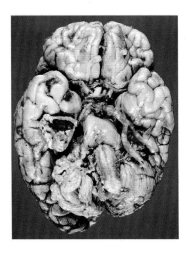

Figure 24.1   Schwannoma classically affecting the cerebellopontine angle of the right hemisphere. (Courtesy of Dr. Stephen Coons, with permission from Barrow Neurological Institute, Phoenix, AZ.)

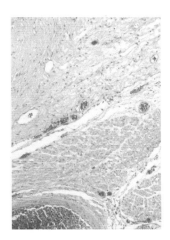

Figure 24.2   The schwannoma (top) is distinct from the associated nerve (below). Schwannomas are only rarely intraneural or invasive. Hematoxylin and eosin (H&E), original magnification 200×. (Courtesy of Dr. Stephen Coons, with permission from Barrow Neurological Institute, Phoenix, AZ.)

Tumors are histologically composed of neoplastic Schwann cells (Figure 24.2) in two basic architectural patterns: Antoni A (Figure 24.3a) with variably but generally highly cellular areas of bland, elongated cells in a dense, diffuse, fibrous stroma and Antoni B (Figure 24.3b) with a less dense, often myxoid stroma. Foci of nuclear palisades, termed

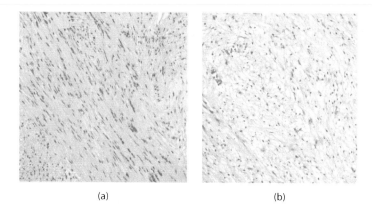

(a)   (b)

Figure 24.3 (a) Antoni A pattern: Hypercellular proliferation of bland spindle cells arranged in intersecting fascicles in diffuse dense fibrous stroma. The slender nuclei demonstrate little atypia. H&E, original magnification 400×. (Courtesy of Dr. Stephen Coons, with permission from Barrow Neurological Institute, Phoenix, AZ.) (b) Antoni B pattern: Similar bland spindle cells but slightly less cellular in a loose edematous/myxoid stroma. H&E, original magnification 400×. (Courtesy of Dr. Stephen Coons, with permission from Barrow Neurological Institute, Phoenix, AZ.)

Figure 24.4 Verocay bodies are pairs of opposing nuclear palisades separated by anucleate fibrillary zones. These are pathognomonic for schwannoma. H&E, original magnification 400×. (Courtesy of Dr. Stephen Coons, with permission from Barrow Neurological Institute, Phoenix, AZ.)

Verocay bodies, are common in the Antoni A regions (Figure 24.4). Inflammatory and lipid-laden cells may also be present in varied quantities. Vasculature in tumors is typically hyalinized and with thickened walls. Nuclei are typically bland but occasional scattered atypical cells are common. In contrast, malignant peripheral nerve sheath tumors (MPNST)

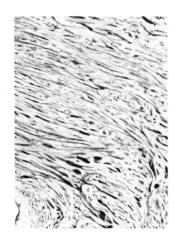

Figure 24.5   Schwannomas demonstrate diffuse strong positivity for S-100. S-100 DAB/hematoxylin, original magnification 200×. (Courtesy of Dr. Stephen Coons, with permission from Barrow Neurological Institute, Phoenix, AZ.)

will have diffuse significant atypia. Mitoses may be absent or generally low in number with a correspondingly low Ki-67 labeling index. Again, MPNST differs by demonstrating high proliferative activity. Schwannomas are strongly and diffusely immunopositive for S-100 (Figure 24.5), distinguishing them from MPNST. Typically, schwannomas will also demonstrate collagen IV positivity and SOX10 expression.

24.7 Treatment

Schwannoma treatment options include surgical removal, radiation, or clinical observation. Conventional schwannomas are typically benign with low risk for malignancy; thus supportive care and observation is often indicated when the tumor is asymptomatic. Tumors must be monitored for growth and subsequent mass effect on nerves and the spinal cord. There is controversy over removal of asymptomatic tumors, considering their high chance for growth. Symptomatic schwannomas and rapidly growing tumors should be removed.

Surgical removal of schwannomas is the primary treatment modality. Minimally invasive microsurgery using ultrasound guidance can be performed on tumors localized near nerves and vessels. For patients who are poor surgical candidates or in whom complete resection is unlikely, stereotactic radiosurgery may be used. Chemotherapy is considered in malignant cases, although highly effective regimens are not well established.

Additionally, combination therapy, namely, surgery with adjuvant radiation, may be considered, particularly in recurrent cases.

24.8 Prognosis

The overall survival of schwannomas is excellent, as most are benign and slow growing with limited ability to infiltrate and metastasize. Vestibular schwannomas are the most common and complete resection is considered curative. The recurrence rate is between 1% and 2% after gross removal in most large cohort studies [3]. Neurological function of the facial nerve is usually preserved, but the great majority of patients have some degree of hearing deficit on the affected side. Larger tumors (>1 cm in diameter) carry a higher risk for total hearing loss and facial nerve weakness.

Spinal schwannomas have an excellent prognosis, and the vast majority can achieve gross total resection with few to no resulting neurological problems. They have a very low rate of recurrence after removal. However, trigeminal nerve and other cranial nerve schwannomas pose more difficulty in achieving complete resection due to either their location or involvement of local tissues.

It is extremely rare for peripheral nerve sheath tumors to become malignant, but when they do occur they carry a significantly worse prognosis. Malignant schwannomas grow rapidly and are more likely to infiltrate tissues, leading to incomplete resection. These aggressive tumors may require chemotherapy or radiation therapy in addition to surgery. Distant metastases may occur and can result in deposits to lung, brain, liver, and bone.

References

1. Bhatheja K, Field J. Schwann cells: Origins and role in axonal maintenance and regeneration. *Int J Biochem Cell Biol.* 2006;38(12):1995–9.
2. Schneider AB, Ron E, Lubin J, et al. Acoustic neuromas following childhood radiation treatment for benign conditions of the head and neck. *Neuro Oncol.* 2008;10(1):73–8.
3. Ahmad RARL, Sivalingam S, Topsakal V, Russo A, Taibah A, Sanna M. Rate of recurrent vestibular schwannoma after total removal via different surgical approaches. *Ann Otol Rhinol Laryngol.* 2012;121(3):156–61.

SECTION III
Tumors of the Meninges

25
Meningioma

25.1 Definition

Affecting the meninges, meningioma (or *meningeal tumor*) comprises 15 histological subtypes/variants that are organized into benign, atypical, and malignant categories [1].

25.1.1 Benign meningioma

Benign meningioma (WHO Grade I, also known as *Grade I meningioma*) is a slow-growing, noninvasive, and noncancerous neoplasm accounting for 65%–80% of all meningiomas. Displaying pleomorphic and occasional mitotic features (<4 mitoses/10 high-power fields [HPF]), benign meningioma is further divided into nine histological subtypes (meningiothelial, fibrous [fibroblastic], transitional [mixed], psammomatous, angiomatous, microcystic, secretory, lymphoplasmacyte-rich, and metaplastic), of which meningiothelial, fibroblastic, and transitional meningiomas are most common. Some benign meningiomas showing two or fewer of the five criteria of greater cellularity (small cell formation with high nuclear-to-cytoplasmic ratio, macronucleoli, sheetlike growth, hypercellularity, and spontaneous necrosis) are designated as benign meningioma with atypical features, which incur a higher risk of recurrence than benign meningioma without atypical features.

25.1.2 Atypical meningioma

Atypical meningioma (WHO II, also known as *Grade II meningioma*) is a fast-growing, moderately invasive neoplasm accounting for 15%–20% of all meningiomas. Showing either high mitotic activity (4–19 mitoses/10 HPF) or three of the five criteria of greater cellularity, atypical meningioma is further divided into chordoid, clear cell, and atypical subtypes.

25.1.3 Malignant (anaplastic) meningioma

Malignant (anaplastic) meningioma (WHO Grade III, also known as *Grade III meningioma*) is a fast-growing, aggressive, and cancerous neoplasm with a tendency to invade the surrounding brain tissue, recur, and metastasize extracranially. Accounting for 1%–4% of meningiomas, malignant meningioma displays ≥20 mitoses/10 HPF and is separated into rhabdoid, papillary, and malignant (anaplastic) subtypes [2].

25.2 Biology

As the membranous coverings of the brain and spinal cord, the meninges are composed of three layers: the dura mater, arachnoid mater, and pia mater. The *dura mater* (or the *pachymeninges*) is the outermost, thick, tough, and inextensible layer situated beneath the bones of the skull and vertebral column. The *arachnoid mater* is the middle layer lying underneath the dura mater. Small projections of the arachnoid mater into the dura (known as *arachnoid granulations*) allow cerebrospinal fluid (CSF) contained in the subarachnoid space to reenter the circulation via the dural venous sinuses. The *pia mater* is the inner layer that lies below the subarachnoid space. As the only covering to follow the contours of the brain (the gyri and fissures), the pia mater is very thin, vascular, and fibrous and adheres tightly to the surface of the brain and spinal cord. Collectively, the arachnoid mater and the pia mater are often referred to as the *leptomeninges*.

Meningioma arises from arachnoid cells or meningiothelial cells (flattened ovoid cells with prominent nuclei) of the arachnoid mater, as well as arachnoidal cells in other places. The tumor is commonly found in the falx (a groove running between the two hemispheres of the brain) and parasagittal (near or close to the superior sagittal sinus at the top of the groove) (25%), convexity (20%), sphenoid wing (also called the *sphenoid ridge*, lying behind the eyes) (20%), olfactory groove (along the nerves that run between the brain and the nose) (10%), suprasellar (above a bony depression that houses the pituitary gland) (10%), posterior fossa (lying on the underside of the brain) (10%), intraventricular (associated with the connected chambers of CSF that circulates throughout the CNS) (2%), intraorbital (around the eye sockets of the skull) (<2%), and spinal (<2%) regions. Multifocal lesions are observed in ~9% of patients on imaging and 16% of patients in autopsy studies.

25.3 Epidemiology

Meningioma is the most common nonglial intracranial tumor and the second most common CNS tumor (after gliomas) in adults. Accounting for >30% of all primary brain tumors, meningioma mainly affects people of 40–70 years of age, with a marked female bias (female-to-male ratio of 2:1

intracranially and 4:1 in the spinal cord), although atypical and malignant meningiomas may be slightly more common in males. There is an association with neurofibromatosis type 2 (NF2) in patients under 40 years of age. Children are seldom involved, as pediatric patients make up only 2.5% of the total meningioma cases [3].

25.4 Pathogenesis

Risk factors for meningioma are radiation exposure, NF2, female gender, oral contraceptives, and hormone replacement therapy.

NF2 is an autosomal dominant disorder characterized by mutations in the *NF2* gene, located on 22q12.2, which contribute to reduced expression of the *NF2* gene product merlin (moesin, ezrin, and radixin-like protein; a member of the 4.1 family of proteins with tumor suppressor function for many different cell types) and to increased levels of Yes-associated protein (YAP). Deletion in the long arm of chromosome 22 (loss of heterozygosity on chromosome 22q) is found in 40%–70% of meningioma cases. Decreased expression of another member of the 4.1 family of proteins, the Dal-1 protein, is also noted in up to 76% of sporadic meningiomas. Combined Dal-1 and Merlin losses are observed in 70% of anaplastic, 60% of atypical, and 50% of benign meningiomas. Other chromosome 22 genes such as PDGF-B (c-sis [22q13.1]), and bcr (BCR [22q11]) may also play a part in the tumorigenesis of meningioma [4].

Additional chromosomal/genetic changes implicated in the pathogenesis of meningioma include chromosomes 1 (TP73 [1p36.3]), 2 (GLI2 [2q14.2]), 7 (SMO [7q32.1]), 8 (c-mos [8q11], SFRP1 [8p11.21], c-myc [8q24]), 9 (CDNK2A [9p21], CDNK2B [9p21], ARF [9p21], PTCH1 [9q22.3]), 10 (PTEN [10p23.3], SUFU [10q24]), 11 (Ha-ras [11p15.5], IGF2 [11p15.5]), 12 (FOXM1 [12p13.3], GLI1 [12q13.3]), 14 (c-fos [14q24.3], AKT1 [14q32.32]), 16 (GLIS2 [16p13.3], CDH1 [16q22.1]), 17 (STAT3 [17q21.2]), and 18 (bcl-2 [18q21.33]) [4].

25.5 Clinical features

Meningioma usually grows inward and compresses nerves originating from or entering into the brain or spinal column (cranial and spinal nerves) and other associated structures, causing reactive swelling in brain tissue surrounding the tumor, blocking the flow of CSF in the brain and spinal cord, and leading to obstructive hydrocephalus. Meningioma may also grow outward toward the skull, causing skull thickening [3].

Clinically, meningioma often manifests as headache, muscle weakness (or paresis) in arms or legs, back pain (spinal meningioma), facial numbness (sphenoid wing meningioma), seizures (cranial meningioma), blurred vision (intraorbital, suprasellar, and olfactory groove meningiomas), hearing loss (posterior fossa meningioma), speech problems, confusion, memory loss (falx meningioma), neurological deficits (convexity meningioma), and personality changes (intraventricular meningioma). Some patients may be asymptomatic.

25.6 Diagnosis

Given the clinical similarity between meningioma and normal aging (e.g., visual changes, loss of memory, menopausal effects, and unsteady gait), imaging techniques (e.g., CT and MRI) and histological investigation are required for confirmation of meningioma.

Meningioma is a very soft to extremely firm, fibrous, or calcified tumor of a light tan color that may appear either as a globose, rounded, bosselated, lobulated, well-circumscribed, smoothly marginated extra-axial mass abutting the dura (similar to a fried egg seen in profile) or en plaque (a carpet-like growth pattern along the dura) with extensive regions of dural thickening (particularly over the sphenoid wing and pterion) on image studies [5].

On CT, meningioma is usually hyperdense, with 20%–25% of cases demonstrating nodular, fine and punctate, or dense calcifications. Occasional invasion into the cranial vault may cause characteristic hyperostosis. Surrounding parenchymal vasogenic edema may be identified in hypodense brain tissue. Less common findings consist of hemorrhage, cyst formation, and necrosis. Atypical and malignant meningiomas are usually larger.

On MRI, meningioma is almost isodense with the gray matter on T1-weighted images and hypointense to hyperintense on T2-weighed images. A crest of CSF (so-called CSF crest) is also seen around the tumor (indicative of the extra-axial location of the tumor) on T2-weighted images.

Contrast-enhanced MRI shows an area of dural enhancement (so-called dural tail) in half of the cases. MR spectroscopy reveals increased alanine (1.3–1.5 ppm), glutamine/glutamate, and choline in cellular tumor but reduced N-acetylaspartate of non-neuronal origin, as well as creatine. These features are helpful in distinguishing meningioma from mimics.

Histologically, benign meningioma subtypes have low mitotic features (<4 mitoses/10 HPF) and do not usually show features of greater cellularity (small cell formation with high nuclear-to-cytoplasmic ratio, macronucleoli, sheetlike growth, hypercellularity, spontaneous necrosis). Among benign meningioma subtypes, *meningothelial (syncytial) meningioma* shows epithelioid, round to polygonal cells arranged in lobules or whorls with moderate amount of amphophilic to eosinophilic cytoplasm. The cells may have intranuclear clear vacuoles and intranuclear pseudoinclusions in addition to fuzzy intercellular border (syncytial appearance). *Transitional (mixed) meningioma* is intermediate between meningothelial and fibroblastic meningiomas, with meningothelial whorls, and psammoma bodies. *Fibroblastic (fibrous) meningioma* contains spindle cells with bland nuclei arranged in fascicles or storiform pattern, together with collagen deposition, and fibrous components. *Psammomatous meningioma* has numerous meningothelial whorls with psammoma bodies (comprising about half of the tumor) at centers. *Angiomatous (vascular) meningioma* shows large peritumoral edema; blood vessels with small- to medium-sized vascular channels and hyalinized walls; the microvascular and macrovascular types; and degenerative nuclear atypia (in combination with microcystic meningioma). Microcystic meningioma contains large peritumoral edema; extracellular microcysts with pale, eosinophilic mucinous fluid; tumor cells with thin, spidery cytoplasmic processes; and numerous pleomorphic cells. Secretory meningioma shows large peritumoral edema; tumor cells with intracellular lumina containing eosinophilic, periodic acid-Schiff (PAS)–positive secretions; intracellular secretory vacuoles (being several times larger that the nuclei); focal epithelial differentiation; positivity for cytokeratin and carcinoembryonic antigen (CEA); and high number of mast cells. Lymphoplasmacyte-rich meningioma displays extensive, chronic inflammatory cell infiltration (obscuring the meningothelial componen); and plasma cells with Russell bodies. Metaplastic meningioma shows metaplastic components (singly or in combination, involving bone, cartilage, myxoid area) and xanthomatous/lipomatous components (involving bone and cartilage); occasional osseous component; and classic meningothelial components.

Atypical meningioma subtypes have moderate mitotic activity (4–19 mitoses/10 HPF) or three of the features of greater cellularity. Among atypical meningioma subtypes, *chordoid meningioma* contains epithelioid cells with bubbly cytoplasm and Alcian blue–positive myxoid stroma; patchy to prominent lymphoplasmacytic infiltration; and chordoid component mixed with meningothelial component. *Clear cell meningioma* (often occurring in the spinal cord and posterior fossa) demonstrates glycogen-rich cytoplasmic clearing but not extracellular secretory vacuoles (by PAS); a pure clear cell tumor with classic meningioma; and dense stromal and perivascular blocky collagen deposition. *Atypical meningioma* shows features of atypical changes in any of the variants, but does not meet the diagnostic criteria of clear cell or chordoid meningioma.

Malignant (anaplastic) meningioma subtypes demonstrate high mitotic activity ($\geq$20 mitoses per 10 HPF). Among malignant meningioma subtypes, *papillary meningioma* (often affecting the posterior fossa) shows a perivascular arrangement of tumor cells, and the papillary or pseudopapillary growth pattern (resulting from dyscohesion of the tumor cells); and tends to arise from more conventional meningiomas. *Rhabdoid meningioma* contains the rhabdoid cells with a discohesive growth pattern and an intracytoplasmic globular, eosinophilic inclusion (displacing the nuclei with vesicular/prominent nucleoli to the periphery); rhabdoid changes superimposed on a papillary meningioma background; and areas with conventional meningiomas. *Malignant (anaplastic) meningioma* shows frank anaplastic changes (similar to metastatic carcinoma, melanoma, and sarcoma) [6].

Immunohistochemically, proliferative markers such as Ki-67 (MIB-1), anti-phosphohistone-H3, and apoptotic index correlate with the biologic behavior in meningioma. Further, meningioma is positive for vimentin, epithelial membrane antigen (EMA, 70%), S-100, androgen, progesterone, and estrogen receptors but negative for cytokeratins (with the exception of the secretory meningioma subtype) [2].

25.7 Treatment

Surgery (craniotomy) represents the primary treatment for meningioma [2]. Radiation therapy (including stereotactic radiosurgery or SRS) may be used for tumors that cannot be removed with surgery, tumors that are not completely removed in surgery, malignant/anaplastic tumors, or recurrent tumors [4]. Radiation can kill malignant tumor cells or shrink the tumor [7]. Chemotherapy (e.g., a combination of cyclophosphamide, doxorubicin, and vincristine; hydroxyurea; interferon-α; and somatostatin analogs, sorafenib, sunitinib, vatalanib) may also be considered for inoperable meningiomas, recurrent disease, and aggressive histologies [8]. Other therapies include hormonal therapies (tamoxifen and flutamide) [5].

25.8 Prognosis

Benign, atypical, and malignant meningiomas have 5-year overall survival rates of 95%, 80%, and 20%, respectively, and 5-year recurrence rates of 3%, 40%, and 78%, respectively. The median overall survival for malignant meningioma is about 2–3 years [5].

References

1. Louis DN, Perry A, Reifenberger G, et al. The 2016 World Health Organization classification of tumors of the central nervous system: A summary. *Acta Neuropathol.* 2016;131(6):803–20.

2. Rogers L, Barani I, Chamberlain M, et al. Meningiomas: Knowledge base, treatment outcomes, and uncertainties. A RANO review. *J Neurosurg.* 2015;122(1):4–23.

3. Gump WC. Meningiomas of the pediatric skull base: A review. *J Neurol Surg B Skull Base.* 2015;76(1):66–73.

4. Miller R Jr, DeCandio ML, Dixon-Mah Y, et al. Molecular targets and treatment of meningioma. *J Neurol Neurosurg.* 2014;1(1). pii:1000101.

5. Bi WL, Zhang M, Wu WW, Mei Y, Dunn IF. Meningioma genomics: Diagnostic, prognostic, and therapeutic applications. *Front Surg.* 2016;3:40.

6. Varlotto J, Flickinger J, Pavelic MT, et al. Distinguishing grade I meningioma from higher grade meningiomas without biopsy. *Oncotarget.* 2015;6(35):38421–8.

7. Walcott BP, Nahed BV, Brastianos PK, Loeffler JS. Radiation treatment for WHO grade II and III meningiomas. *Front Oncol.* 2013;3:227.

8. Kaur G, Sayegh ET, Larson A, et al. Adjuvant radiotherapy for atypical and malignant meningiomas: A systematic review. *Neuro Oncol.* 2014;16(5):628–36.

26
Melanocytic Tumors

26.1 Definition

Arising from the melanocytes within the leptomeninges, primary melanocytic tumors consist of meningeal melanocytosis, melanomatosis, melanocytoma, and melanoma [1].

26.1.1 Meningeal melanocytosis

Meningeal melanocytosis (or *diffuse melanocytosis/melanosis*) is a diffuse infiltration of the subarachnoid space of the brain and spinal cord by melanocytes, possibly associated with neurocutaneous melanosis. The latter is a phakomatosis caused by congenital dysplasia of the neuroectodermal melanocyte precursor cells, leading to excessive (focal or diffuse) proliferation of melanin-producing cells in the skin and leptomeninges [2].

26.1.2 Meningeal melanomatosis

Meningeal melanomatosis (or *primary leptomeningeal melanomatosis*) is the malignant form of diffuse involvement resulting from the spread of malignant melanocytes into the leptomeninges and Virchow–Robin spaces with superficial invasion of the brain. Also referred to as a meningeal variant of primary malignant melanoma, meningeal melanomatosis is a rare, aggressive neoplasm of the central nervous system (CNS) with a poor prognosis [3].

26.1.3 Meningeal melanocytoma

Meningeal melanocytoma (formerly *melanotic meningioma*) is a rare, pigmented, slow-growing, and typically benign neoplasm that may sometimes show malignant behavior. Similar to meningeal melanoma, meningeal melanocytoma is not associated with pigmented lesions elsewhere, including benign congenital pigmented nevi or frank cutaneous malignant melanoma [4].

26.1.4 Meningeal melanoma

Meningeal melanoma (or *primary malignant melanoma*) is a malignant tumor with a worse prognosis than meningeal melanocytoma [5].

26.2 Biology

Melanocytes are melanin-producing cells that originate from the neural crest during early embryogenesis and that develop mainly in the skin but occasionally in the eye, mucous membranes, or the CNS. As normal, yet sparse, cells of the leptomeninges within the CNS, melanocytes cover the base of brain, brainstem, and spinal cord. Consequently, the areas most commonly affected by primary intracranial melanocytic tumors are the pons, cerebellum, cerebral peduncles, medulla, interpeduncular fossa, and inferior surfaces of the frontal, temporal, and occipital lobes.

Although both meningeal melanocytosis and meningeal melanomatosis involve the supra- and infratentorial leptomeninges and the superficial brain parenchyma (particularly in the cerebellum, brainstem, and temporal lobes), only meningeal melanomatosis spreads the malignant melanocytes from the leptomeninges to the Virchow–Robin spaces, and superficially within the brain substance. Nonetheless, both lesions generally occur in the setting of dermatologic syndromes (e.g., neurocutaneous melanosis syndrome and nevus of Ota). Similarly, while both meningeal melanocytoma and meningeal melanoma may affect any area of the meninges, they show a predilection for the spinal cord, posterior fossa and Meckel's cave. Meningeal melanocytoma is a benign lesion, whereas primary malignant melanoma of the leptomeninges is malignant. However, neither of these lesions is associated with pigmented lesions elsewhere (e.g., benign congenital pigmented nevi or frank cutaneous malignant melanoma) [6].

26.3 Epidemiology

Primary melanocytic tumors are rare diseases accounting for <1% of brain tumors. Meningeal melanocytosis often occurs in children of <2 years, and meningeal melanomatosis predominantly affects adults (with peak prevalence in the fourth decade). Meningeal melanocytoma arises at any age with a slight predilection for females. Meningeal melanoma most frequently involves adults of 15–71 years (mean age of 50 years), particularly men, and represents approximately 1% of all melanoma cases.

26.4 Pathogenesis

Melanocytic tumors of the CNS demonstrate overlapping histological characteristics but show distinct biological behavior. Melanocytomas harbor *GNAQ/11* mutations and copy number variants (CNVs) involving chromosomes 3 and 6, whereas melanomas frequently contain mutation in the

TERT promoter and additional oncogene mutations, together with recurrent chromosomal losses involving chromosomes 9, 10, and 6q, as well as gains of 22q. In contrast, melanotic schwannomas possess a complex karyotype with recurrent monosomy of chromosome 22q and variable whole chromosomal gains and recurrent losses commonly involving chromosomes 1, 17p, and 21, in addition to mutations in the *BRAF V600* or *KIT* genes. Like uveal melanoma, primary leptomeningeal melanocytic neoplasms appear to have oncogenic mutations in *GNAQ* and *GNA11* but not in *BRAF, NRAS,* and *HRAS.* Oncogenic mutations in primary melanocytic tumors are remarkably different between adult patients (e.g., *GNAQ, GNA11*) and children (e.g., *NRAS*) [7].

Chromosomal and genetic changes mentioned above and still uncovered play an important role in melanocytic tumor genesis, and subsequent tumor growth and infiltration exert mechanical, physiological, and immunological impact on the leptomeninges and other affected tissues, contributing to various clinical manifestations (e.g., headache, seizures, intracranial hypertension, ataxia, and neuropsychiatric signs) [7].

26.5 Clinical features

Clinically, meningeal melanocytosis may cause hydrocephalus, mass effect, and neuropsychiatric symptoms, which resemble those of neurocutaneous melanosis and congenital nevi. Meningeal melanomatosis is associated with headache, nausea/vomiting, seizures, monoparesis, cranial nerve palsies, spinal cord compression, and psychiatric disturbance. Meningeal melanocytomas may induce headache, seizures, hydrocephalus, chronic basal meningitis, multiple cranial nerve palsies, chronic spinal arachnoiditis, psychiatric disturbances, stillbirth, intracranial hemorrhage of the meninges or subdural space, myelopathy, and radiculopathy. Similarly, meningeal melanoma may show mass effect/cord compression symptoms [8].

26.6 Diagnosis

Primary melanocytic neoplasms of the CNS often present as either discrete masses or diffuse lesions that are recognized as meningeal (diffuse) melanocytosis, melanomatosis, melanocytoma, or (malignant) melanoma. Thorough physical examination is helpful in discriminating melanocytic tumors (e.g., meningeal melanocytoma) from a primary cutaneous, ocular, or mucosal melanoma. CT and MRI are useful for preliminary identification of melanocytic tumors. Further histological, biochemical, and molecular characterization of biopsy and CSF samples is critical in achieving correct premortem diagnosis of melanocytic tumors.

Meningeal melanocytosis is a diffuse lesion that may render the leptomeninges over the base of the brain diffusely thickened and embedded in a dark tissue. In some cases, a small, dark nodule is found within the dorsal column of the spinal cord. On MRI, the tumor may be isointense or hyperintense on T1-weighted images (due to the paramagnetic properties of the melanin) and hypointense on T2-weighted images. Histologically, the tumor shows abundance of uniform cells (spindled, round, oval, or cuboidal) with a lack of malignant features and without brain invasion [8].

Meningeal melanomatosis may present with bilateral/diffuse slow abnormalities without epileptic activity. Brain MRI reveals focal or diffuse nodular enhancement of the leptomeninges and cranial nerves. Spine MRI shows nodular contrast enhancement of the meninges, conus, and cauda. Whole-body Fluorine 18 fluorodeoxyglucose positron emission tomography/computed tomography (F-18 FDG PET/CT) helps rule out the extracranial origin of melanoma lesions.

Meningeal melanocytoma is usually a solitary, well-demarcated mass with black or red-brown discoloration. Histologically, the tumor is a solitary, circumscribed lesion, without invading adjacent structures. It contains nests (reminiscent of whorls) of relatively uniform cells with melanin pigment, oval nuclei, and eosinophilic nucleoli; and shows no more than 1 mitoses per 10 high power fields [8].

Malignant melanoma is a solitary, well-demarcated mass of a black or red-brown color. Histologically, the tumor displays hypercellular sheets or nests of spindled or epithelioid cells with prominent nucleoli. Invasion of adjacent structures or necrosis is observed along with atypical mitosis (5 per 10 high power fields).

Immunohistologically, melanocytic tumors are positive for human melanoma black 45 (HMB-45), Melan A, MAGE1, CD63, vimentin, S-100 MITF, and Ki-67 (<2% for melanocytomas; about 8% for melanomas). Melanocytomas and melanomas are known to be negative for glial fibrillary acidic protein (GFAP), keratins, and epithelial membrane antigen (EMA) [8].

Differential diagnoses for primary melanocytic tumors include melanotic schwannoma, metastatic melanoma, melanocytic neuroectodermal tumors, neurocutaneous melanosis, pigmented meningioma, and melanoblastosis, among others.

Meningeal melanocytoma is distinguishable histologically from primary malignant melanoma on the basis of mitotic activity, cytological atypia, necrosis, and invasion of the adjacent structures. Compared to malignant melanoma, meningeal melanocytoma lacks mitotic activity, nuclear pleo-morphism, and hyperchromaticity, as well as an indolent growth pattern spanning >4 years. Unlike meningioma, meningeal melancytoma rarely displays tumor calcification and hyperostosis of the adjacent bone at CT. However, MRI does not reliably distinguish meningeal melanocytomas from other extraaxial neoplasms, such as meningiomas, schwannomas, and malignant melanoma, which occur in similar locations.

26.7 Treatment

Treatment options for primary melanocytic tumors consist of gross total resection and adjuvant chemoradiation therapy. Complete surgical resec-tion of meningeal melanocytoma may be curative, although in up to 22% of cases, lesions recur within 3 years after surgery. Radiation therapy is preferred for cases in which intraoperative hemorrhage may be severe or a recurrence takes place. Hydrocephalus in meningeal melanocytoma may be treated with placement of a ventriculoperitoneal shunt, and addition of a filter to the apparatus helps prevent spreading in the rare event of malignant transfor-mation. Similarly, complete surgical excision is usually possible with localized leptomeningeal melanomas and represents the best treatment option [9,10].

26.8 Prognosis

Prognosis for meningeal melanocytosis is relatively poor even with histolog-ically benign cases, given its potential to develop focal melanomatosis and metastasize to distant sites. Meningeal melanocytoma generally has a good prognosis, with 5-year overall survival rates of 83% after complete resec-tion and 40% in incomplete resection in the absence of radiation therapy. Application of radiation therapy increases 5-year overall survival in subtotal resection to 91%–92%. Malignant melanoma also has a poor prognosis (6 years in spine) but is certainly better than metastatic melanoma in the CNS (6 months).

References

1. Louis DN, Perry A, Reifenberger G, et al. The 2016 World Health Organization classification of tumors of the central nervous system: A summary. *Acta Neuropathol.* 2016;131:803–20.
2. Painter TJ, Chaljub G, Sethi R, Singh H, Gelman B. Intracranial and intraspinal meningeal melanocytosis. *Am J Neuroradiol.* 2000;21(7):1349–53.

3. Berzero G, Diamanti L, Di Stefano AL, et al. Meningeal melanomatosis: A challenge for timely diagnosis. *Biomed Res Int*. 2015;2015:948497.

4. Reddy R, Krishna V, Sahu BP, Uppin M, Sundaram C. Multifocal spinal meningeal melanocytoma: An illustrated case review. *Turk Neurosurg*. 2012;22(6):791–4.

5. Smith AB, Rushing EJ, Smirniotopoulos JG. Pigmented lesions of the central nervous system: Radiologic-pathologic correlation. *Radiographics*. 2009;29(5):1503–24.

6. Trinh V, Medina-Flores R, Taylor CL, Yonas H, Chohan MO. Primary melanocytic tumors of the central nervous system: Report of two cases and review of literature. *Surg Neurol Int*. 2014;5:147.

7. Küsters-Vandevelde HV, Küsters B, van Engen-van Grunsven AC, Groenen PJ, Wesseling P, Blokx WA. Primary melanocytic tumors of the central nervous system: A review with focus on molecular aspects. *Brain Pathol*. 2015;25(2):209–26.

8. PathologyOutlines.com website. Melanocytic tumors/melanoma. http://www.pathologyoutlines.com/topic/cnstumormelanocytictumor.html. Accessed December 15, 2016.

9. Jaiswal S, Vij M, Tungria A, Jaiswal AK, Srivastava AK, Behari S. Primary melanocytic tumors of the central nervous system: A neuroradiological and clinicopathological study of five cases and brief review of literature. *Neurol India*. 2011;59(3):413–19.

10. Cockerell CJ, Tschen J, Evans B, et al. The influence of a gene expression signature on the diagnosis and recommended treatment of melanocytic tumors by dermatopathologists. *Medicine (Baltimore)*. 2016;95(40):e4887.

27

Hemangiopericytoma and Hemangioblastoma

27.1 Definition

Arising from fibroblasts and considered to be within the spectrum of solitary fibrous tumors of the dura, hemangiopericytoma (HPC) may appear as (i) true HPC (including central nervous system [CNS] HPC, spleen HPC, myopericytoma, infantile myofibromatosis, and sinonasal HPC; with clear evidence of myoid or pericyte differentiation); (ii) solitary fibrous tumors, and (iii) non-HPC tumor (e.g., synovial sarcoma, with occasional display of HPC-like features). In the 2016 WHO classification of CNS tumors, true HPC and solitary fibrous tumor (SFT) are restructured as one entity (HPC/SFT) and a grading system is adapted to accommodate this change.

Consisting of HPC (WHO Grade II) and anaplastic HPC (WHO Grade III), CNS/meningeal HPC is clinicopathologically distinct from meningioma (a more common, benign tumor of the meninges, WHO Grade I), based on its tendency to invade locally, metastasize extraneurally, and recur frequently (up to 92% of cases in 15 years) [1].

CNS hemangioblastoma (WHO Grade I) is the prototypic lesion of von Hippel–Lindau (VHL) syndrome, involving the cerebellum (80%) and spinal cord (20%). Although considered histologically benign, CNS hemangioblastoma in the forms of peritumoral cysts and solid tumors may cause a multitude of symptoms and even death [2].

27.2 Biology

The meninges are the coverings of the brain composed of three layers of connective tissue membranes (dura, arachnoid, and pia mater). The dura mater (or *pachymeninx*, *pachy* meaning "thick") is the outer layer, the arachnoid mater (*arachn* meaning "spider") the middle layer, and the pia mater (*pia* meaning "tender") the inner layer. The arachnoid mater and pia mater are together referred to as the *leptomeninges* (*lepto* meaning "thin"). Cerebrospinal fluid (CSF) is found in the subarachnoid space between the arachnoid mater and the pia mater. By anchoring the brain to the surrounding bones, the meninges provide stability and cushioning for the CNS and participate in the physiological and pathological activities of the CNS as well.

The most common cell type of connective tissue is fibroblast, which is derived from primitive mesenchyme and involved in the synthesis of the extracellular matrix, the intermediate filament protein vimentin (a marker of mesodermal origin), and collagen. Fibroblasts are present in various parts of the body, including the meninges. In the inner pia layer, meningeal fibroblasts organize and maintain the basement membrane, which is a critical attachment point for radial glial endfeet. CNS HPC (or meningeal HPC) evolves from fibroblasts located in the meninges. The most common meningeal HPCs are intracranial/intradural (particularly extramedullary and occasionally intramedullary).

The dura mater and pia mater in the meninges are highly vascularized, with blood vessels perforating through the membranes to supply the underlying neural tissue. Arising from the vascular tissues of the meninges (brain and spinal cord) and other parts (kidney, retina, adrenal glands, and pancreas) of the body, hemangioblastoma is generally a slow-growing tumor associated with VHL syndrome.

27.3 Epidemiology

CNS HPC is a rare, aggressive tumor of the meninges, accounting for 2.5% of all meningeal tumors and 1% of intracranial tumors. The tumor is typically detected in adults of the third and fourth decades (mean age of 44 years), and only about 10% of cases involve children. It shows a slight male predilection (male-to-female ratio of 1.4:1) [3,4].

Representing <2% of CNS tumors, CNS hemangioblastoma usually occurs in adults, although that associated with VHL syndrome may develop at much younger ages (mean age at diagnosis of 25 years). Men and women are equally susceptible.

27.4 Pathogenesis

Multiple genetic changes (e.g., activation of the oncogenes Bcl-2, p53, and Ki-67 and the insulin-growth factor [IGF] family) are implicated in the pathogenesis of CNS HPC. IGF ligands together with insulin-like growth factor I receptor (IGF IR) may contribute to hypoglycemia and subsequent phenotype transformation (including the formation of characteristic thin-walled, branching blood vessels, or *staghorn vessels*) [3].

CNS hemangioblastoma is a common and important manifestation of VHL syndrome. Inherited in an autosomal dominant manner, about 80% of VHL patients have an affected parent and about 20% patients have a *de novo* pathogenic variant. *VHL* is a tumor suppressor gene located on the short arm of chromosome 3 (3p) that encodes a protein (pVHL) forming part of the E3 ubiquitin ligase and is involved in proteasomal degradation. The *VHL* gene is prone to germline mutation (95%) and somatic inactivation (5%), leading to four VHL phenotypes (1, 2A, 2B, and 2C). Inactivation (loss of heterozygosity) of the wild-type *VHL* allele leads to tumorigenesis in target organs, including the viscera (kidneys, pancreas, adrenal glands, and adnexal organs) and the CNS [5].

Four VHL phenotypes (1, 2A, 2B, and 2C) have been identified. VHL type 1 has a low risk for pheochromocytoma, with common occurrence of retinal angioma, CNS hemangioblastoma, renal cell carcinoma, pancreatic cysts, and neuroendocrine tumors. VHL type 2 has a high risk for pheochromocytoma in addition to retinal angiomas and CNS hemangioblastoma. VHL type 2 is further subdivided into three categories: Type 2A (commonly associated with pheochromocytoma, retinal angiomas, and CNS hemangioblastoma, but not with renal cell carcinoma), Type 2B (commonly associated with pheochromocytoma, retinal angioma, CNS hemangioblastomas, pancreatic cysts, and neuroendocrine tumor, and a high risk for renal carcinoma), and Type 2C (commonly associated with pheochromocytoma only) [6].

Potential risk factors for VHL syndrome include tobacco smoking, exposure to chemicals and industrial toxins, and contact sports (if adrenal or pancreatic lesions are present).

27.5 Clinical features

The relatively large size and frequent supratentorial location of CNS/meningeal HPC produce a mass effect (intracranial hypertension), leading to headache, vomiting, seizures, and focal neurological dysfunction. In anaplastic HPC, systemic metastases to the liver, lung, and bone may induce other related symptoms. The average duration of HPC symptoms before diagnosis is about 3–6 months [4].

Intracranial hemangioblastomas (87%–97%; predominantly in the posterior fossa and occasionally supratentorially) are associated with headache, vomiting, gait disturbances, or ataxia. Spinal hemangioblastomas (3%–13%; intradural, commonly occurring in the cervical or thoracic regions, and occasionally involving the entire cord) often present with pain and sensory and motor loss (due to cord compression). Polycythemia may occur in ~20% of CNS hemangioblastoma cases due to erythropoietin production.

27.6 Diagnosis

27.6.1 CNS/meningeal HPC

CNS/meningeal HPC is a solitary, lobulated mass usually present in a supratentorial location. The tumor is highly vascular and has a tendency to erode adjacent bone [7].

On CT, the tumor often shows vivid enhancement and erosion of adjacent bone. On MRI, the tumor appears as isointense to gray matter on T1 and T2; and reveals vivid enhancement, heterogeneous signal, a narrow base of dural attachment, and a dural tail sign (more common in Grade II tumors) on T1 C+ (Gd). Although meningeal HPC and meningioma appear similar on both CT and MRI, meningioma mainly occurs in older patients (>50 years), is smoother, has central spoke-wheel vascular supply, does not erode adjacent bone, tends to cause hyperostosis, does not metastasize, and usually has a broad dural attachment and dural tail (both of which are absent in HPC). In contrast, HPC typically shows the existence of large vessels within the lesion and brain edema along with adjacent bone erosion.

Histologically, meningeal HPC demonstrates a densely cellular neoplasm comprised spindle cells arranged in a solid and storiform pattern, with the background blood vessels having a distinct staghorn-like pattern. Immunohistologically, HPC cells are diffusely positive for vimentin (85%) and CD34 (30%–100%) but negative for S-100, SMA, and CD99. Differential diagnoses for meningeal HPC include meningioma (EMA+, whorls, psammoma bodies), anaplastic meningioma (EMA+, CD99+/-, chromosomal deletions), fibrous meningioma (80% EMA+, 80% S100+), and glioma, etc.

27.6.2 CNS hemangioblastoma

CNS hemangioblastoma is a solid or cystic tumor of simple cystic, macrocystic, or microcystic form. In 60% of cases, the tumor is a sharply demarcated homogeneous mass composed of a cyst with nonenhancing walls, except for a mural nodule (commonly abutting the pia) with vivid enhancement and prominent serpentine flow voids. In the remaining 40% of cases, the tumor is solid with no cystic cavity [8].

On CT, the mural nodule is isodense to brain with fluid density surrounding the cyst; bright enhancement of the nodule is demonstrated with contrast. On MRI, the tumor appears as a hypointense to isointense mural nodule with CSF signaling cyst content on T1; as a hyperintense mural nodule, with the presence of flow voids (due to enlarged vessels at the periphery of the cyst in 60%–70% of cases), and fluid-filled cyst on T2; the mural nodule appears vividly enhanced on T1 C+ (Gd) [8].

Microscopically, hemangioblastoma shows proliferation of capillaries with variably sized, closely packed, thin-walled vessels containing large neoplastic stromal cells with pink to clear foamy cytoplasm and fine vacuoles containing PAS+ lipid. Other features consist of hyperchromatic nuclei, cyst wall with gliosis and Rosenthal fibers (resembling pilocytic astrocytoma), numerous mast cells in the tumor mass, and the absence of atypia, fibrillar cells, necrosis, and mitotic figures. Immunohistochemically, stroma is positive for NSE, lipid (at frozen section), reticulin, CD34, VEGF, inhibin alpha and occasionally positive for erythropoietin, GFAP, and S-100 but negative for epithelial membrane antigen, keratin, and CD10.

Identification of the *VHL* gene mutation is helpful for the confirmation of CNS hemangioblastoma. Indeed, detection of a heterozygous germline *VHL* pathogenic variant (1, 2A, 2B, or 2C) by molecular genetic testing establishes the diagnosis even in cases that clinical and radiographic features are inconclusive [5,6].

27.7 Treatment

Gross total resection and adjuvant radiotherapy are recommended treatments for CNS HPC. The range of radiation doses applied for the initial treatment after the surgery is between 50 and 60 Gy (with 1.8-2.0 Gy/day). Chemotherapy is largely ineffective for CNS HPC at the present [9].

Surgery and radiation therapy (most frequently stereotactic radiosurgery or SRS) are useful for treatment of CNS hemangioblastoma associated with VHL syndrome. SRS is particularly relevant for CNS hemangioblastoma patients who cannot tolerate surgical resection and have non–surgically resectable lesions [9,10].

27.8 Prognosis

CNS HPC has 5-, 10-, and 15-year overall survival rates of 85%, 68%, and 43%, respectively. With a tendency to recur even after gross total resection, CNS HPC shows 5- and 10-year overall disease recurrence-free rates of 65% and 76%, respectively. Intradural tumors carry a better prognosis than extradural spinal tumors.

Early treatment and diligent surveillance are crucial for increasing life expectancy in patients with CNS hemangioblastoma, which currently stands at 49 years [10].

References

1. Noh SH, Lim JJ, Cho KG. Intracranial hemangiopericytomas: A retrospective study of 15 patients with a special review of recurrence. *J Korean Neurosurg Soc.* 2015;58(3):211–16.

2. Kinyas S, Ozal SA, Guclu H, Gurlu V, Esgin H, Gurkan H. Von Hippel-Lindau disease: The clinical manifestations and genetic analysis results of two cases from a single family. *Balkan J Med Genet.* 2016;18(2):65–70.

3. Hall WA, Ali AN, Gullett N, et al. Comparing central nervous system (CNS) and extra-CNS hemangiopericytomas in the surveillance, epidemiology, and end results program: Analysis of 655 patients and review of current literature. *Cancer.* 2012;118(21):5331–8.

4. Das A, Singh PK, Suri V, Sable MN, Sharma BS. Spinal hemangiopericytoma: An institutional experience and review of literature. *Eur Spine J.* 2015;24(Suppl 4):S606–13.

5. Muscarella LA, la Torre A, Faienza A, et al. Molecular dissection of the VHL gene in solitary capillary hemangioblastoma of the central nervous system. *J Neuropathol Exp Neurol.* 2014;73(1):50–8.

6. Frantzen C, Klasson TD, Links TP, Giles RH. Von Hippel-Lindau Syndrome. In: Pagon RA, Adam MP, Ardinger HH, et al. (editors). GeneReviews® [Internet]. Seattle (WA): University of Washington, Seattle; 1993-2016. 2000 May 17 [updated 2015 Aug 6].

7. Radiopedia.org. Meningeal haemangiopericytoma. https://radiopaedia.org/articles/meningeal-haemangiopericytoma. Accessed December 20, 2016.

8. Medscape. Brain imaging in hemangioblastoma. http://emedicine.medscape.com/article/340994-overview#a4. Accessed December 20, 2016.

9. Shirzadi A, Drazin D, Gates M, et al. Surgical management of primary spinal hemangiopericytomas: An institutional case series and review of the literature. *Eur Spine J.* 2013;22(Suppl 3):S450–9.

10. Mills SA, Oh MC, Rutkowski MJ, Sughrue ME, Barani IJ, Parsa AT. Supratentorial hemangioblastoma: clinical features, prognosis, and predictive value of location for von Hippel-Lindau disease. *Neuro Oncol.* 2012;14(8):1097–104.

SECTION IV
Tumors of the Sellar Region

28
Craniopharyngioma

28.1 Definition

Craniopharyngioma is a rare, benign tumor of the sellar region that consists of two recognized histopathologic variants: adamantinomatous and papillary [1].

Adamantinomatous (ordinary) craniopharyngioma (ACP, WHO Grade I) is characterized by epithelium that grows in cords, lobules, and whorls, including both palisading peripheral columnar epithelium and loosely arranged epithelium (called *stellate reticulum*), with "wet" keratin as a hallmark. Being less solid than papillary craniopharyngioma (PCP, WHO Grade I), ACP affects both children and adults.

PCP is characterized by well-differentiated monomorphic squamous epithelium covering fibrovascular cores with thin capillary blood vessels and scattered immune cells including macrophages and neutrophils. Due to a lack of surface maturation in the epithelium, no wet keratin is observed. As a solid intracranial tumor, PCP causes high levels of morbidity almost exclusively in adults.

Recent genetic testing reveals the existence of at least four subtypes within craniopharyngioma: (i) ACP CTNNB1 mutated (96% with activating CTNNB1 mutation, β-catenin nuclear stain), (ii) ACP CTNNB1 wild type, (iii) PCP BRAF V600E mutated (95% with BRAF V600F mutation), and (iv) PCP BRAF wild type [2]. This opens another potential avenue for craniopharyngioma treatment with targeted agents.

28.2 Biology

The sellar region is the area around the sella turcica, which is a saddle-shaped, bony depression (or hollow) within the sphenoid bone at the skull base, and in which the pituitary gland is situated. Above the sellar region lies the suprasellar cistern (or chiasmatic cistern), with several vital structures traversing the area (e.g., the circle of Willis, optic nerves, optic chiasm, hypothalamus, pituitary stalk, and the infundibular and suprachiasmatic recesses of the third ventricle).

Probably developing from the small nests of cells near the pituitary stalk in the suprasellar region, craniopharyngioma (also known as *Rathke pouch tumor*, *hypophyseal duct tumor*, or *adamantinoma*) is an epithelial–squamous, calcified, and cystic neoplasm, which may extend anteriorly into the prechiasmatic cistern and subfrontal spaces; posteriorly into the prepontine and interpeduncular cisterns, cerebellopontine angle, third ventricle, posterior fossa, and foramen magnum; and laterally toward the subtemporal spaces. Due to its critical location, craniopharyngioma often compresses or infiltrates the vital neurological areas and affects many functions of the brain such as hormone making, growth, and vision.

28.3 Epidemiology

Craniopharyngioma has an estimated incidence of 0.5–2 cases per million per year and accounts for 2%–5% of all primary brain tumors. Whereas the adamantinous histological type (with cyst formation) is mainly seen in children and adolescents of 5–14 years and represents 30%–50% of cases, the papillary histological type occurs in adults over 45 years (peak incidence between 50 and 75 years) and represents 50%–70% of cases [3]. ACP shows a slight female predilection, and has the potential to transform into malignant tumor.

28.4 Pathogenesis

Over 90% of ACP contain mutations in exon 3 of the *CTNNB1* gene. As exon 3 encodes the degradation targeting box of beta-catenin, a mutation in exon 3 results in nuclear accumulation of beta-catenin protein and dysregulation of the Wnt signaling pathway [2,4]. On the other hand, >90% of PCP harbor *BRAF V600E* mutations. It is of interest to note that *CTNNB1* and *BRAF* alterations are mutually exclusive, clonal, and specific to each subtype [5].

28.5 Clinical features

Despite its slow growth, craniopharyngioma can put pressure on the brain, the optic chiasm, and the pituitary gland, leading to clinical symptoms such as headache (due to blocked third ventricle and obstructive hydrocephalus), visual impairment (bitemporal hemianopia), endocrine deficits (due to hypothalamus damage), balance disorder, dry skin, fatigue, fever, hypersomnia, lethargy, myxedema, nausea, vomiting, short stature, polydipsia, polyuria (diabetes insipidus), and postsurgical weight gain.

In children, clinical manifestations of craniopharyngioma are similar, including headache, nausea or vomiting, vision changes, loss of balance or trouble walking, increase in thirst or urination, unusual sleepiness or change in energy level, changes in personality or behavior, short stature or slow growth, hearing loss, and weight gain.

28.6 Diagnosis

Diagnosis of craniopharyngioma involves physical exam, medical history review, neurological exam (for changes in mental status, coordination, senses, and reflexes), visual field exam (for changes in central vision and peripheral vision), CT and MRI, blood chemistry, and hormone studies and histologic confirmation (which is generally required before treatment).

Macroscopically, ACP is a lobulated, partly cystic mass (often >5 cm), which shows irregular interface with adjacent brain and contains viscous, proteinaceous, dark greenish-brown fluid (resembling motor oil). The characteristic white speckled appearance of "wet" keratin nodules is frequently visible. On the other hand, PCP is a discrete, encapsulated mass without the complex multicystic architecture and fluid-filled spaces upon sectioning. Further, if cystic, PCP may contain clear liquid instead of viscous, oily fluid.

On CT, ACP shows prominent peripheral calcification. On MRI, it displays a complex solid/cystic lesion with heterogeneous signal intensity. The cysts containing fluid of high protein content appear hyperintense on T1-weighted images. Solid areas reveal enhancement after gadolinium administration. The combination of solid, cystic, and calcified tumor components is diagnostic for ACP. In contrast, PCP shows an enhancing, predominantly solid, circumscribed mass without the calcification or complex cystic architecture, but the papillary architecture is sometimes evident on MRI [6].

Histologically, ACP shows tongues of tumor extending into hypothalamic parenchyma, and contains compact sheets, nodules and trabeculae of squamous epithelium, which is composed of loose plump central cells (stellate reticulum) and nuclear palisading peripheral cells. Other features include nodules of plump, anucleate squames (ghost cells) and "wet" keratin, intralobular whorl-like formations, cystic degeneration, calcification, and xanthogranulomatous reaction; cyst fluid–containing cholesterol crystals, cholesterol clefts, and reactive giant cells; variable necrosis, inflammation, and Rosenthal fibers. Immunohistochemically, ACP is positive for CK7, CK8, CK19 and beta-catenin (in keratin whorls); but negative for mutant BRAF (V600E) [6,7].

On the other hand, PCP is a solid lesion that typically shows papillary configuration with cauliflower-like morphology, sheets of well differentiated nonkeratinizing squamous epithelium, crude papillae around fibrovascular cores, and small collagenous whorls. Nodules of wet keratin, stellate reticulum, and calcification are absent. If cystic, the fluid does not resemble motor oil. Immunohistochemically, PCP is positive for CK7, EMA, mutant BRAF (V600E), and beta-catenin (membranous only, with nuclei and cytoplasm being negative); but negative for CK8 and CK20. Differential diagnoses for PCP include ACP (irregular/infiltrative borders, complex architecture, "wet" keratin, calcification, peripheral palisading, loose "stellate reticulum," beta-catenin in keratin whorls, positivity for CK8) and Rathke's cleft cysts (usually cystic without solid component, squamous epithelium with ciliated or mucus containing cells, positivity for CK8 and CK20) [6,8].

Molecularly, ACP contains a mutation in exon 3 of the beta-catenin gene (*CTNNB1*), whereas-PCP does not exhibit this mutation but harbors a BRAF v600E mutation, which is absent in ACP. BRAF VE1 immunohistochemistry is valuable for differentiation of PCP from ACP and possibly also Rathke's cleft cysts.

28.7 Treatment

Treatment options for craniopharyngioma range from surgical resection to external-beam radiation therapy, stereotactic radiosurgery, intracavitary radiation therapy, intracavitary chemotherapy or biologic therapy, intravenous biologic therapy, and cyst drainage [9,10].

For ACP, the standard care is gross total excision or subtotal resection followed by radiation therapy. For PAP, its smoother surface facilitates easier gross total resection than ACP.

Tumor resection, shunt operation, and stereotactic or open implantation of an intracystic catheter help restore normal cerebrospinal fluid flow, reduce the volume and pressure of the cyst, and instill sclerosing substances (bleomycin).

Inhibition of beta-catenin provides a useful approach for the targeted therapy of PCP with *BRAF V600E* mutation.

Chemotherapy with dabrafenib/trametinib contributes to reduction of the Ki-67 proliferation index from over 20% in the pretreatment tumor to less than 0.5% in the on-treatment tumor.

Side effects from surgery and radiation include physical problems (e.g., seizures); behavior problems, changes in mood/feelings/thinking/learning/memory, and second cancers.

28.8 Prognosis

Prognosis for craniopharyngioma is dependent on the size and location of the tumor, amount of residual tumor, and patient age. Craniopharyngioma is generally a benign tumor with high overall survival rates (95% at 3 years, 91% at 5 years, and 87% at 10 years). Prognostically, PCP appears to fair marginally better than ACP. However, the quality of life may be impaired in long-term survivors as a result of sequelae caused by the anatomical proximity of craniopharyngioma to the optic nerve, pituitary gland, and hypothalamus [10].

References

1. Lubuulwa J, Lei T. Pathological and topographical classification of craniopharyngiomas: A literature review. *J Neurol Surg Rep.* 2016;77(3):e121–7.
2. Apps JR, Martinez-Barbera JP. Molecular pathology of adamantinomatous craniopharyngioma: Review and opportunities for practice. *Neurosurg Focus.* 2016;41(6):E4.
3. Muller HL. Childhood craniopharyngioma. Recent advances in diagnosis, treatment and follow-up. *Horm Res.* 2008;69(4):193–202.
4. Martinez-Barbera JP. Molecular and cellular pathogenesis of adamantinomatous craniopharyngioma. *Neuropathol Appl Neurobiol.* 2015;41(6):721–32.
5. Brastianos PK, Santagata S. Endocrine tumors: BRAF V600E mutations in papillary craniopharyngioma. *Eur J Endocrinol.* 2016;174(4):R139–44.
6. Medscape. Craniopharyngioma pathology. http://emedicine.medscape.com/article/1744011-overview#a6. Accessed January 31, 2017.
7. PathologyOutlines.com website. Adamantinomatous craniopharyngioma. http://www.pathologyoutlines.com/topic/cnstumoradamcraniopharyngioma.html. Accessed January 31, 2017.
8. PathologyOutlines.com website. Papillary craniopharyngioma. http://www.pathologyoutlines.com/topic/cnstumorpapcraniopharyngioma.html. Accessed January 31, 2017.
9. Trippel M, Nikkhah G. Stereotactic neurosurgical treatment options for craniopharyngioma. *Front Endocrinol (Lausanne).* 2012;3:63.
10. PDQ Pediatric Treatment Editorial Board. *Childhood Craniopharyngioma Treatment (PDQ®): Health Professional Version.* PDQ Cancer Information Summaries. Bethesda, MD: National Cancer Institute (US), 2002.

Glossary

anaplasia: A term used to describe cancer cells with a total lack of differentiation and with resemblance to original cells either in functions or structures or both; also known as *dedifferentiation* (backward differentiation).

aneuploidy: the presence of abnormal number of chromosomes in a cell (see *diploidy*).

angiogenesis: the growth of new capillary blood vessels from pre-existing vessels in the body (especially around a developing neoplasm).

apoptosis: Programmed cell death, with cells that are damaged beyond repair typically dying as they swell and burst, spilling their contents over their neighbors.

atypia: The state of being not typical or normal. In medicine, atypia is an abnormality in cells in tissue, which may or may not be a precancerous indication associated with later malignancy.

benign tumor: A slow-growing, noncancerous tumor that does not invade nearby tissue or spread to other parts of the body. In most cases, a benign tumor has a favorable outcome, with or without surgical removal. However, a benign tumor in vital structures such as nerves and blood vessels, or that is undergoing malignant transformation, often has serious consequences (see hamartoma).

biopsy: A procedure to remove tumor tissue or cells or tissues for microscopic examination. This is usually conducted through excisional biopsy (removal of an entire lump of tissue), incisional biopsy (removal of part of a lump or a sample of tissue), core biopsy (removal of tissue using a wide needle), or fine needle aspiration biopsy (removal of tissue or fluid using a thin needle).

calcification: The accumulation of calcium salts (e.g., calcium phosphate) in body tissues such as tumor mass, where they do not usually appear. This leads to tissue hardening and produces a dense opacity on a radiographic image.

cancer (plural *cancers* or *cancer*): A group of diseases involving the uncontrolled expansion of abnormal cells that have the potential to invade nearby tissue and/or spread to other parts of the body via the bloodstream or lymphatic vessels (see *tumor, neoplasm, lesion*).

carcinoma: A type of cancer that begins in a tissue (called *epithelium*) that lines the inner or outer surfaces of the body.

CSF (cerebrospinal fluid): A clear, colorless body fluid produced from arterial blood by the choroid plexuses of the lateral and fourth ventricles of the brain. The main functions of CSF are to act as a cushion for the brain's cortex and to provide basic mechanical and immunological protection to the brain inside the skull.

CT (computerized tomography): Also known as *computed tomography scan, CT scan*, or *computerized axial tomography* (CAT), CT utilizes an X-ray machine linked to a computer together with a dye injected into a vein or swallowed to take a series of detailed pictures of affected organs or tissues in the body from different angles, in order to determine the precise location and dimensions of a tumor.

cyst: A closed capsule or sac-like structure usually filled with liquid, semi-solid, or gaseous material (but not pus, which is considered an abscess). As an abnormal formation, a cyst on the skin, mucous membranes, or inside palpable organs can be felt as a lump or bump, which may be painless or painful. Whereas cysts due to infectious causes are preventable, those due to genetic and other causes are not. Most cysts are benign (noncancerous).

dedifferentiation: See *anaplasia*.

desmoplasia: The growth of fibrous or connective tissue around a neoplasm, causing dense fibrosis; it is considered a hallmark of invasion and malignancy.

differentiated: A term used to describe how much or how little tumor tissue looks like the normal tissue it came from. Well-differentiated cancer cells look more like normal cells and tend to grow and spread more slowly than poorly differentiated or undifferentiated cancer cells (see undifferentiated).

diploidy: the presence of normal number (two sets) of chromosomes in a cell (see aneuploidy).

dysplasia: The overgrowth of immature cells at the location where the number of mature cells is decreasing. This term is particularly used for when cellular abnormality is restricted to the new tissues.

EMA (epithelial membrane antigen): EMA is expressed in a range of normal tissues (nearly all apical surfaces of ductular and glandular epithelium, normal plasma cells, renal distal convoluted tubules, renal collecting ducts, and perineural cells) and neoplasms (e.g., meningioma, epithelioid sarcoma, and synovial sarcoma but not schwannoma).

endoscopy: A thin, tubelike instrument with a light and a lens for checking for abnormal areas inside the body.

FISH (fluorescence *in situ* hybridization): FISH is used for determining the positions of particular genes, for identifying chromosomal abnormalities, and for mapping genes of interest.

GFAP (glial fibrillary acidic protein): An intermediate filament protein of 52 kDa found in glial cells such as astrocytes and ependymal cells.

grade, grading: A measure of cell anaplasia (reversion of differentiation) in tumor, it is based on the resemblance of the tumor to the tissue of origin. Depending on the amount of abnormality, a tumor is graded as 1 (well differentiated; low grade), 2 (moderately differentiated; intermediate grade), 3 (poorly differentiated; high grade), or 4 (undifferentiated; high grade) (see *stage*, TNM).

H&E stain: Combined use of hematoxylin (positively charged) and eosin (negative charged) to stain nucleic acids (negatively charged) in blue and amino groups in proteins (negatively charged) in pink, respectively.

hamartoma: a benign, tumor-like, focal malformation resulting possibly from a developmental error. Composed of an abnormal/disorganized mixture of cells and tissues, hamartoma grows at the same rate as the surrounding tissues and rarely invades or compresses surrounding structures significantly. In contrast, a true benign tumor may grow faster than surrounding tissues and compress nearby structures. Despite its benign histology, hamartoma may be implicated in some rare but life-threatening clinical issues such as those associated with neurofibromatosis type I and tuberous sclerosis. A nonneoplastic mass (e.g., Lhermitte–Duclos disease, hemangioma) can also arise in this way, contributing to misdiagnosis (see *benign tumor*).

hydrocephalus: An abnormal accumulation of CSF in the ventricles of the brain, which can compress and damage the brain.

hyperplasia: A disease associated with the increase in number of normal-looking cells, leading to an enlarged organ; it is also known as *hypergenesis*.

IHC (immunohistochemistry): A technique that exploits the principle of antibodies binding specifically to antigens in biological tissues to visualize the distribution and localization of specific cellular components within cells and in the proper tissue context.

Ki-67: A nuclear protein (also known as KI-67 or MKI67) associated with cellular proliferation and ribosomal RNA transcription. Ki-67 protein is present during all active phases of the cell cycle [G1 (pre-DNA synthesis), S (DNA synthesis), G2 (post-synthesis), and M (mitosis)] but is absent in resting phase (G_0). A monoclonal antibody (MIB-1) raised against Ki-67 protein allows accurate immunohistochemical detection of active or growing cells. The fraction of Ki-67-positive tumor phase (the *Ki-67 labeling index* or MIB-1 labeling index) often correlate to the aggressiveness and thus the clinical course of cancer.

lesion: A term in medicine to describe all abnormal biological tissue changes, such as a cut, a burn, a wound, or a tumor. In cancer, *lesion* is used interchangeably with *tumor*, *cancer*, or *neoplasm* (see *cancer, tumor, neoplasm*).

LOH (loss of heterozygosity): A gross chromosomal event that results in loss of the entire gene and the surrounding chromosomal region.

malignancy: The state or presence of a malignant tumor.

malignant tumor: A tumor with the capability of invading surrounding tissues, producing metastases, and recurring after attempted removal.

metaplasia: The reversible replacement of one differentiated cell type with another mature differentiated cell type.

MIB-1: see Ki-67.

mitotic figure, mitosis: Microscopic detection of the chromosomes as tangled, dark-staining threads in cells undergoing mitosis; it is often expressed as mitotic figures per high power fields (hpf, usually a 400-fold magnification) (mitotic activity index) or per 1,000 tumor cells (mitotic index). As mitotic cell count per 10 hpf equals to an area 0.183 mm^2, the American Joint Committee on Cancer specifies that mitotic rate (the proportion of cells in a tissue undergoing mitosis) be reported as mitoses per mm^2, with a conversion factor of 1 mm^2 equaling 4 hpf.

MRI (magnetic resonance imaging): Also called *nuclear magnetic resonance imaging* (NMRI), MRI relies on a magnet, radio waves, and a computer to take a series of detailed pictures of affected areas inside the body that help pinpoint the location and dimension of tumor mass, if present. MRI has better image resolution than CT. It includes T1-weighted, T2-weighted, fluid-attenuated inversion recovery (FLAIR, also called dark fluid technique), and diffusion-weighted imaging (DWI) sequences. T1-weighted images reveal anatomical details and information about venous sinus permeability or pathologic blush (e.g., water and CSF appear dark; fat and calcification appear white/gray). Use of intravenous contrast gadolinium in T1-weighted sequences further enhances and improves the quality of the images. T2-weighted images provide information about edema, arteries and sinus permeability (e.g., water appears white/hyperintense; fat and calcification appear gray/dark). FLAIR sequences remove the effects of fluid (which normally covers a lesion) from the resulting images (e.g., CSF appears dark; edema appears enhanced). DWI sequences help visualize acute infarction and other inflammatory lesions.

mutation: A change in the structure of a gene caused by the alteration of single base units in DNA or the deletion, insertion, or rearrangement of larger sections of genes or chromosomes, leading to the formation of a variant that may be transmitted to subsequent generations.

necrosis: A form of cell injury leading to the premature/unprogrammed death of cells and living tissue caused by autolysis (due to too little blood flowing to the tissue; see *apoptosis*).

neoplasia: A term that describes abnormal growth or proliferation of cells, resulting in a tumor that can be cancerous.

neoplasm: A new and abnormal growth of tissue in a part of the body; it is used interchangeably with *tumor* or *cancer*.

oncogene: A gene whose mutation or abnormally high expression can transform a normal cell into a tumor cell.

parenchyma: The functional tissue of an organ as distinguished from the connective and supporting tissue (see *stroma*).

pleomorphism: A term used in histology and cytopathology to describe variability in the size, shape, and staining of cells and/or their nuclei; it is a feature characteristic of malignant neoplasms and dysplasia.

PCR (polymerase chain reaction): A procedure for rapid, in vitro production of multiple copies of particular DNA sequences relevant to diagnosis.

PET (positron emission tomography): A PET scan combines a computer-based scan with a radioactive glucose (sugar) injected into a vein to generate a rotating picture of the affected area, with malignant tumor cells showing up more brightly due to their more active uptake of glucose than normal cells.

radiotherapy: Also called *radiation*, *radiation therapy*, or *X-ray therapy*, radiotherapy involves delivery of radiation externally through the skin or internally (brachytherapy) for destruction of cancer cells or inhibition of their growth.

radiography: A term used to collectively describe electromagnetic radiation (especially X-ray) based procedures to visualize the internal structure of a non-uniformly composed and opaque object such as the human body (see *MRI*, *CT*).

stage, staging: Referring to the size of tumor and/or the extent of cancer cells that have spread in the body, the stage of a solid tumor is often expressed clinically as stage 0, I, II, III, IV using information on the pathological stage of a tumor obtained with the TNM system (see *grade*, *TNM*).

stroma: The parts of a tissue or organ that have a connective and structural role and that do not conduct the specific functions of the organ (e.g., connective tissue, blood vessels, nerves, ducts) (see *parenchyma*).

TNM: A system (also known the *TNM system*) designed by American Joint Committee on Cancer to pathologically stage a solid tumor. *T* (*tumor*; TX, T0, T1, T2, T3, T4) indicates the depth of the tumor invasion—the higher the number (between 0 and 4, with x being unknown or not assessed), the further the cancer has spread. *N* (*nodes*; NX, N0, N1, N2, N3) indicates whether the lymph nodes are affected—a number between 0 and 3 describes how much the cancer has spread to lymph nodes near the bladder. *M* (*metastasis*; MX, M0, M1) indicates if the cancer has spread to other parts of the body. Thus, the pathological stage of a given tumor may be designated as T1N0MX or T3N1M0 (with numbers after each letter providing more details about the tumor/cancer). Knowing a tumor's pathological stage helps select the most appropriate treatments and offers a useful guidance on its prognosis (see *grade, stage*).

tumor: A swelling of a part of the body, generally without inflammation, caused by an abnormal growth of tissue, either benign or malignant (see *cancer, neoplasm, lesion*).

tumor suppressor gene: A gene (also known as *antioncogene*) that regulates cell division, repairs DNA mistakes, or instructs cells when to die. When a tumor suppressor gene is mutated, cell growth gets out of control (see *apoptosis*).

ultrasound: A device for delivering sound waves that bounce off tissues inside the body like an echo and recording the echoes to create a picture (sonogram) of areas inside the body.

undifferentiated: The presence of very immature and primitive cells that do not look like cells in the tissue of their origin. Undifferentiated cells are said to be anaplastic and an undifferentiated cancer is more malignant than a cancer of that type that is well differentiated (see *anaplasia, differentiated*).

Index

A

ACP. *See* Adamantinomatous (ordinary) craniopharyngioma
ACPP. *See* atypical choroid plexus papilloma
Adamantinoma, 184
Adamantinomatous (ordinary) craniopharyngioma (ACP), 183, 185, 186, 187
Afferent neurons, 78
Aging population, 1
AJCC. *See* American Joint Commission on Cancer
American Joint Commission on Cancer (AJCC), 3
Anaplasia, 2
Anaplastic astrocytoma, 9, 12–13
Anaplastic ependymoma, 62
Anaplastic ganglioglioma, 83
Anaplastic pleomorphic xanthoastrocytoma, 9
Angiocentric glioma, 9, 31–36
    biology, 32
    clinical features, 33
    definition, 31
    diagnosis, 34
    epidemiology, 32
    pathogenesis, 32
    prognosis, 35
    treatment, 35
Angiomatous (vascular) meningioma, 165
Apoptosis, 1
Arachnoid granulations, 162
Astroblastoma, 9, 37–40
    biology, 37
    clinical features, 38
    definition, 37
    diagnosis, 38–39
    epidemiology, 37
    pathogenesis, 37
    prognosis, 39
    treatment, 39
Astrocytes, 10
Astrocytoma, 9–15, 75
    biology, 10
    clinical features, 11
    definition, 9
    diagnosis, 12–13
    epidemiology, 10–11
    pathogenesis, 11
    prognosis, 14
    treatment, 14
Astroglia, 10
AT/RT. *See* Atypical teratoid/rhabdoid tumor
Atypical choroid plexus papilloma (ACPP), 65, 67, 69
Atypical meningioma, 161, 165

Atypical teratoid/rhabdoid tumor (AT/RT), 131, 133, 134, 136
Axons, 78

B

Benign meningioma, 161
Benign tumors, 1
Biopsy, 4–5
Bipolar neurons, 78
Brachytherapy, 5

C

Cancer
    metastatic, 1–2
    secondary, 1–2
    staging, 2–4
Cancer Genome Atlas, 6
Carcinoid tumor, 2
Carcinoma, 2
Central nervous system (CNS), 2–3, 147
Central nervous system (CNS) cancer, 2, 3; *see also specific types*
Central neurocytoma (CN), 89–94
Cerebellar liponeurocytoma (CLPN), 89, 90, 91–93
Cerebral ganglioneuroblastoma, 133
Cerebral neuroblastoma, 133
Cerebral neurocytomas, 89
Cerebrospinal fluid (CSF), 175
Chemotherapy, 5
Chordoid glioma of the third ventricle, 9, 31–36
    biology, 31–32
    clinical features, 33
    definition, 31
    diagnosis, 33–34
    epidemiology, 32
    pathogenesis, 32–33
    prognosis, 35
    treatment, 34–35
Chordoid meningioma, 165
Choroid plexus carcinoma (CPC), 65, 67–68
Choroid plexus papilloma (CPP), 65, 66, 67, 69
Choroid plexus tumors, 65–70
    biology, 65–66
    clinical features, 67
    definition, 65
    diagnosis, 67–68
    epidemiology, 66
    pathogenesis, 66
    prognosis, 69
    treatment, 68–69
Clear cell meningioma, 165

POCKET GUIDES TO
BIOMEDICAL SCIENCES

Series Editor
Dongyou Liu